Quality of meat and fat in pigs as affected by genetics and nutrition

I0038150

The EAAP series is published under the direction of Dr. P. Rafai

EAAP - European Association for Animal Production

ETH *Eidgenössische*
Technische Hochschule
Zürich

ETH Zürich - Eidgenössische Technische Hochschule Zürich

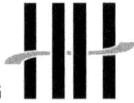

HERMANN HERZER STIFTUNG   FONDATION HERMANN HERZER

Hermann Herzer Foundation

The European Association for Animal Production wishes to express its appreciation to the *Ministero per le Politiche Agricole* and the *Associazione Italiana Allevatori* for their valuable support of its activities

Quality of meat and fat in pigs as affected by genetics and nutrition

Proceedings of the joint session of the EAAP commissions on pig production, animal genetics and animal nutrition

EAAP Publication No. 100
Zurich, Switzerland
25 August 1999

Editors:

Caspar Wenk, José A. Fernández and Monique Dupuis

Wageningen Academic
Publishers

This work is subject to copyright. All rights
are reserved, whether the whole or part of
the material is concerned. Nothing from this
publication may be translated, reproduced,
stored in a computerised system or
published in any form or in any manner,
including electronic, mechanical,
reprographic or photographic, without prior
written permission from the publisher,
Wageningen Academic Publishers, P.O. Box
220, 6700 AE Wageningen, the Netherlands,
www.WageningenAcademic.com

Subject headings:
Pig
Pork quality
Genetics, nutrition, management

The individual contributions in this
publication and any liabilities arising from
them remain the responsibility of the
authors.

ISBN: 978-90-74134-74-3
e-ISBN: 978-90-8686-504-8
DOI: 10.3920/978-90-8686-504-8

ISSN 0071-2477

The designations employed and the
presentation of material in this publication
do not imply the expression of any opinion
whatsoever on the part of the European
Association for Animal Production
concerning the legal status of any country,
territory, city or area or of its authorities,
or concerning the delimitation of its
frontiers or boundaries.

First published, 2000

© Wageningen Academic Publishers
The Netherlands, 2005

The publisher is not responsible for possible
damages, which could be a result of content
derived from this publication.

Content

Foreword 5

Meat consumption and meat quality - How important is it? 7
    C. Wenk

What is pork quality? 15
    H. J. Andersen

Influence of genetics on pork quality 27
    A.G. de Vries, L. Faucitano, A. Sosnicki & G.S. Plastow

Genetic parameters of meat quality traits recorded on Large White and French
Landrace station-tested pigs in France 37
    T. Tribout & J.P. Bidanel

Dissection of genetic background underlying meat quality traits in swine 43
    J. Szyda, E. Grindflek, Z. Liu & S. Lien

Skeletal muscle fibres as factors for pork quality 47
    A. H. Karlsson, R. E. Klont & X. Fernandez

Selection progress of intramuscular fat in Swiss pig production 69
    D. Schwörer, A. Hofer, D. Lorenz & A. Rebsamen

Nutritional and genetic influences on meat and fat quality in pigs 73
    K. Nürnberg, G. Kuhn, U. Küchenmeister & K. Ender

Food waste products in diets for growing-finishing pigs 81
    N. P. Kjos & M. Øverland

Conservation and development of the Bísaro pig. Characterisation and zootechnical
evaluation of the breed for production and genetic management 85
    J. Santos e Silva, J. Ferreira-Cardoso, A. Bernardo & J. S. Pires da Costa

Meat consumption and cancer risk 93
    M. Zimmermann

Who eats meat: factors effecting pork consumption in Europe and the United States 101
    W. Jamison

Breed effect on meat quality of Belgian Landrace, Duroc and their reciprocal crossbred pigs 111
    G. Michalska, J. Nowachowicz, B. Rak & W. Kapelanski

Genetic and energy effects on pig meat quality 115
    Z. Gajic & V. Isakov

Genetic parameters for fattening traits in the Belgian Piétrain population 119
    D. Geysen, S. Janssens & W. Vandepitte

Genetic parameters for colour traits and pH and correlations to production traits 123
    S. Andersen & B. Pedersen

Halothane gene effect on carcass and meat quality by use of Duroc x Piétrain Boars 129
    H. Busk, A. Karlsson & S.H. Hertel

In vivo and post mortem changes of muscle phosphorus metabolites in pigs of different Malignant
Hyperthermia Genotype 135
    M. Henning, U. Baulain, G. Kohn & R. Lahucky

Interactive effects of the HAL and RN major genes on carcass quality traits in pigs:
preliminary results 139
    P. Le Roy, C. Moreno, J.M. Elsen, J.C. Caritez, Y. Billon, H. Lagant, A. Talmant,
    P. Vernin, Y. Amigues, P. Sellier & G. Monin

Meat quantity to meat quality relationships when the RYR1 gene effect is eliminated 143
    J. Kortz, W. Kapelanski, S. Grajewska, J. Kuryl M. Bocian & A. Rybarczyk

Correlations between growth rate, slaughter yield and meat quality traits after the
elimination of the RYR1 gene effect 147
    W. Kapelanski, J. Kortz, J. Kuryl, T. Karamucki & M. Bocian

Effect of the RYR 1 gene on meat quality in pigs of Large White, Landrace and Czech
Meat Pig breeds...151
    R. Bečková & P. David
Development of a highly accurate DNA-test for the *RN* gene in the pig....................157
    C. Looft, D. Milan, J.T. Jeon, S. Paul, C. Rogel-Gaillard, V. Rey, A. Tornsten,
    N. Reinsch, M. Yerle, V. Amarger, A. Robic, E. Kalm, P. Chardon & L. Andersson
Performances of the Piétrain ReHal, the new stress negative Piétrain line......................161
    P. L. Leroy & V. Verleyen
Effect of the RN⁻ gene on the growth rate carcass and meat quality in crossbreeding of
Large White sows with P-76 boars...165
    M. Koćwin-Podsiadła, W. Przybylski, E. Krzęcio, A. Zybert &, S. Kaczorek
Breed and slaughter weight effects on meat quality traits in hal- pig populations.............171
    X. Puigvert, J. Tibau, J. Soler, M. Gispert & A. Diestre
Genotypic and allelic frequencies of the RYRI locus in the Manchado de Jabugo pig breed 175
    A.M. Ramos, J.V. Delgado, T. Rangel-Figueiredo, C. Barba, J. Matos & M. Cumbreras
Intramuscular fat content in some native German pig breeds 181
    U. Baulain, P. Köhler, E. Kallweit & W. Brade
Fatty acid composition and cholesterol content of the fat of pigs of various genotypes.....................185
    J. Csapó, F. Húsvéth, Zs. Csapó-Kiss, P. Horn, Z. Házas & É. Varga-Visi
The effect of paternal breed on meat quality of progeny of Hampshire, Duroc and
Polish Large White boars 189
    J. Nowachowicz, G. Michalska, B. Rak & W. Kapelanski
Comparison of several pig breeds in fattening and meat quality in some experimental
conditions of a Czech region 193
    T. Adamec, B. Naděje, J. Laštovková & M. Koucký
Fat deposition and distribution in three genetic lines of pigs from 10 to 105 kg liveweight 199
    K. Kolstad
Fat score, an index value for fat quality in pigs – its ability to predict properties of backfat
differing in fatty acid composition 203
    K. R. Gläser, M.R.L. Scheeder & C. Wenk
Meat quality with reference to EUROP carcass grading system.............................207
    W. Kapelanski, B. Rak, J. Kapelanska & H. Zurawski
The correlations between the fattening and slaughter performance in pigs 213
    A. Pietruszka, R. Czarnecki & E. Jacyno
Comparison of fat supplements of different fatty acid profile with growing-finishing swine 217
    J. Gundel, A. Hermán Ms., M. Szelényi Ms, & G. Agárdi
The pork meat quality in pigs with a different intensity of nitrogen substances retention..................221
    M. Čechová, V. Prokop, K. Dřímalová, Z. Tvrdoň & V. Mikule
The effect of dietary Ca-fatty acid salts of linseed oil on cholesterol content in
longissimus dorsi muscle of finishing pigs 225
    T. Barowicz
Soybean oil, sex, slaughter weight, cross-breeding - influence on fattening performance and
carcass traits of pigs 229
    R. Kratz, E. Schulz, G. Flachowsky & P. Glodek
Transfer of vitamin E supplements from feed into pig tissues...............................233
    G. Flachowsky, H. Rosenbauer, A. Berk, H. Vemmer & R. Daenicke
Effect of Comfrey (symhytum peregrinum) fed to pigs on meat quality traits 237
    G. Bee, G. J. Seewer Lötscher & P.-A. Dufey
Author index..243
Subject index 245

Foreword

This publication contains the contributions presented in the joint session on "Quality of Meat and Fat as Affected by Genetics and Nutrition" at the 50th Annual Meeting of the European Association for Animal Production (EAAP), Zurich, August 25th 1999. This session was a joint activity of the Commissions on Pig Production, Animal Nutrition and Animal Genetics and was excellently organised and conducted by Prof. C. Wenk.

More than 200 attendants and 37 offered short papers demonstrated the actuality of the topic. In the light of increasing consumer demands for products, which are safe, healthy and wholesome to eat and of consistently high quality, the session was structured to describe meat and fat quality by 6 invited reviews, which further to considerations about manipulation of meat quality by genetical and nutritional means, also included an updated definition of the quality of fresh meat and of convenience food, an appraisal of the effect of meat consumption on human health and lastly an account of public perception and attitudes upon pork and pork production.

The Commissions wish to thank all invited speakers and authors of offered papers for their valuable contributions to this publication. We are particularly indebted to Prof. C. Wenk whom, with characteristic enthusiasm and diligence, also is the main responsible for the publication of these proceedings.

This publication gives an excellent overview of the topics related to the quality of meat. We hope that these proceedings will further stimulate and inspire future research.

Many sincere thanks to the Hermann Herzer Foundation, Switzerland, who sponsored the publication of these proceedings.

José A. Fernández   Ynze van der Honing   Johan van Arendonk
Commission on Pig Production   Commission on Animal   Commission on Animal
                               Nutrition   Genetics

Meat consumption and meat quality - How important is it?

C. Wenk

Institute for Animal Sciences, ETH Zürich

The various ways in which human food is produced are intensively discussed and questioned in modern societies. We expect food from plants and farm animals to be inexpensive, healthy and of course, high quality. In addition, arguments come primarily from biological farming organisations, environmental pressure groups and consumers organisations. In general, we all expect our food to be as natural as possible and free of any toxic, unhealthy or undesired substances.

In highly developed countries where all possible foods are available in any quantity and at all times, we do not always consider the impact of the steady growth of world population. Today there are almost 6 billion inhabitants and in twenty five years' time there are expected to be almost 9 billion inhabitants (FAOSTAT, 1998) on earth requiring food. The goal to produce sufficient food for everybody can only be achieved if the world food production increases by about 2 % per year. Today, more than 800 million people suffer from hunger. Furthermore, the actual world cereal reserves can supply current needs for only a few months (FAO, 1999).

It is expected that world animal production will follow this trend of increasing world population. According to FAOSTAT (1998) world production for pigs and chicken will grow by 1.8 and 2.0 % respectively, in the next 20 years. For beef production a small reduction of 0.4 % is expected, mainly in developed countries.

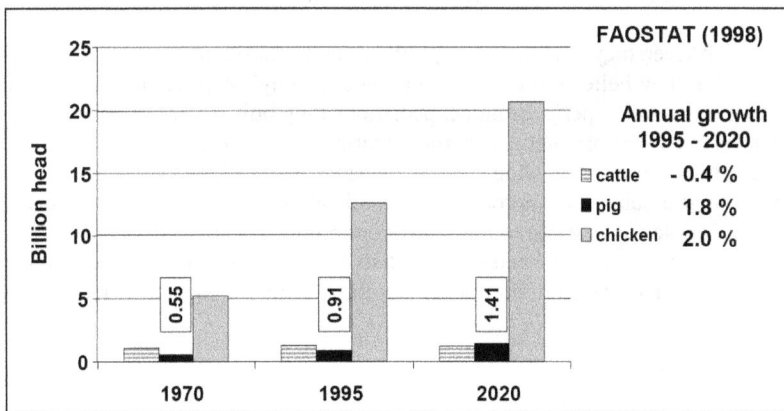

Figure 1.   *World-wide Development of Farm Animal Production.*

Currently, the total world pig population is estimated to be about 1 billion. In twenty years' time, estimates suggest, this will increase to about 1.4 billion pigs, all of which have to be kept under acceptable housing conditions, with good health conditions and fed with adequate feedstuff.

There is no doubt that today's world-wide agriculture productivity must be increased.

However, the consumers in highly developed countries make greater and greater demands on the quality, and idealistic images of food focus attention on issues other than supply or yield; for example the use of new technologies, such as genetic engineering in food and animal feed production, is questioned in many countries world-wide.

We are well aware that decisions on nutrition policy are strongly influenced by idealistic and emotional arguments conveyed by the media and politicians who lack sufficient scientific knowledge in animal production and especially in product quality. Scientists in the fields of nutrition, hygiene or toxicology, on the other hand, try to explain phenomena or calculate the benefits and risks of nutritional variables by first understanding the fundamental physiological or metabolic facts involved. They clearly see nutrition from the "bottom up - view", as shown in Figure 2.

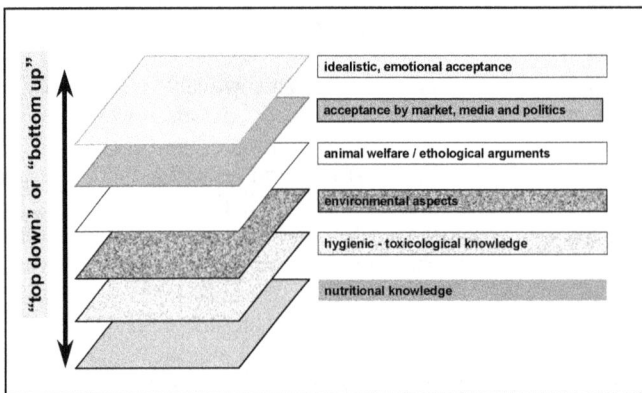

Figure 2.   Decision Making Processes when Evaluating a Food.

The end consumers (whoever they are) with an idealistic and emotional view of nutrition, select food based on what they believe to be as "natural and healthy" as possible and the best for them. In addition to personal experience and expectations they build their opinions mainly on the basis of advertisements or product appearance. During recent years politicians, media and food distributors have stimulated wide-ranging debates on food security and human nutrition. In contrast to the scientific approach, they look at issues from the "top down" perspective (Figure 2). Problems arise due to these different points of view because frequently the consumers with the "top down view" and the scientists with the "bottom up view" do not speak the same "*language*" and therefore do not respect the considerations and arguments of each other.

Pig meat production trends

Pork is the by far the most important meat product globally, with a high protein content. Current annual per capita consumption is about 16 kilograms. In developed countries almost three times more pork is consumed than in developing countries. The almost linear increase of consumption over the last 20 years is mainly based on the intensification of production in Asian countries.

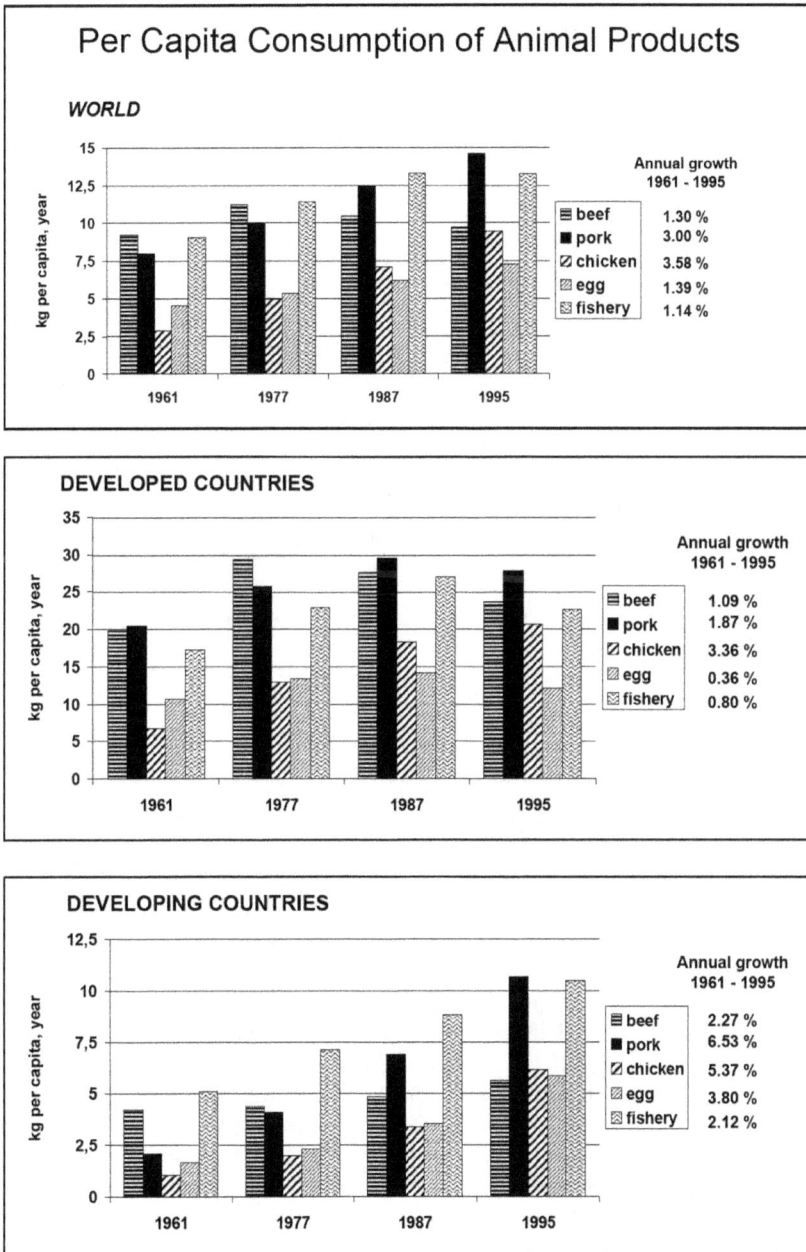

Figure 3. Consumption of Animal Products World-wide, in Developed and Developing Countries.

In developed countries maximum meat consumption could be observed in the early 1980's. Since then there has been a trend of decreasing consumption, mostly of beef and pork. On the other hand, in most countries world wide, chicken consumption is still growing. The main reasons for the decrease of beef and pork consumption are that meat consumption is no longer so closely related with well being and with income in our societies. Furthermore, the consumers have far more possibilities in choosing between different meat cuts and, of course, alternative foodstuffs. When they decide to eat meat they can choose between pork, beef, veal or any other meat that is normally available in any quantity and at any time.

Figure 4.   Meat Consumption in Switzerland (kg edible meat per person 1949 - 1996 (GSF, 1997).

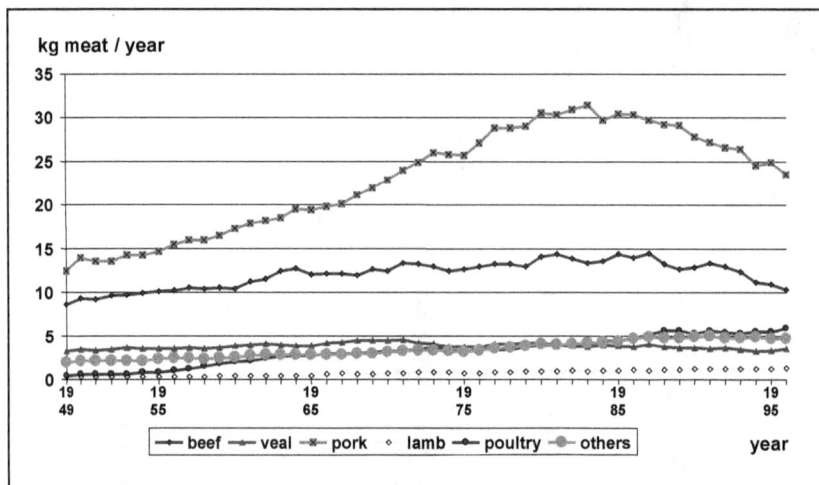

A similar trend concerning meat consumption can also be observed since the Second World War in many European countries and information for Switzerland, a highly developed country, is presented in Figure 4.

Maximal pork consumption was achieved in 1982. Since then there has been an almost linear decline in both pork and beef consumption. During that period scandals, such as BSE, or the misuse of anti-microbial agents, did not specifically affect the trend of reduced meat consumption. However, there are some indications that the negative trend will stop and that a recovery can be expected. Today the daily consumption of pork amounts to approximately 55 g/day. This corresponds to about half of the total meat consumption from all slaughtered animals (beef, veal, sheep and goat, horse, deer etc.). With the consumption of chicken, fish and shellfish, total meat consumption amounts to about 125 g per daily. Compared to many other developed countries, meat consumption in Switzerland is rather moderate. Nevertheless it contributes significantly to the intake of essential nutrients and plays an important and ever increasing role in meeting our requirements of several vitamins (mainly vitamins B1, B2, B6, B12) and trace elements (mainly Fe, Zn or Se).

In several experiments Leonhardt (1998) and Leonhardt & Wenk (1999) have studied the contribution of the mean daily meat consumption in meeting the nutrient requirements of adults in Switzerland according to the recommendations of the German Society of Nutrition (DGE, 1995). From the main meat cuts (pork, beef, veal and chicken) samples were randomly bought in butcheries and analysed for iron, zinc, thiamine, riboflavin and tocopherols. Besides the essential amino acids, meat contributed significantly to the supply of the trace elements.

The requirement of absorbed iron was met in the range of 10-30% and the requirement of absorbed zinc was covered to 30-55%. Red meat cuts, such as beef and pork shoulder are important sources of the trace elements. Pork has a high content of vitamin B1 (thiamine) and it contributed to more than 25% of the requirement of that vitamin. Vitamin B2 (riboflavin) is mainly found in the pork shoulder and chicken thigh. The average lean meat consumption contributed to more than 10% of the requirement. Finally, in Switzerland chicken thigh is a good source of vitamin E, because the chicken feed is often supplemented with this vitamin. However, the consumption of chicken is comparably low. The authors of these studies concluded that even a modest amount of meat, as consumed in Switzerland, is an important source of different vitamins and trace elements. A regular meat consumption in reasonable quantities has therefore a very beneficial effect on the health status of the population.

With the increased production and consumption of pork, the importance of product quality is also increased. Besides high quantities of pork at low prices, attributes such as nutrient content, good sensory quality, shelf-life and resistance against deterioration during processing and storage, are of growing importance in modern food production, particularly for convenience food.

Consumers in the different parts of the world can have very divergent perspectives of the quality criteria of pork. But they decide whether they will buy pork or an alternative product. In the context of the above, the question of quality becomes a relevant issue. Only with the consideration of all factors can a general view be achieved. Depending on the perspective and requirements, the approaches can be very different.

The main quality arguments for products of muscle and adipose tissues are:

- high content of essential nutrients
- low incidence of pale, soft and exudative (PSE) as well as dark, firm and dry (DFD) meat
- high content of intramuscular fat
- good distribution of muscle fibre and connective tissues
- low amount of total body fat, but high fat content in the adipose tissues (no 'empty fat tissues')
- high oxidative stability
- good consistency of the adipose tissues

The priority of these criteria depends on the product utilisation. For instance, for fresh meat that is eaten within a few days other criteria are relevant compared to cured meat products, which are often stored at ambient temperatures over periods as long as one year or even longer (e.g. Iberian ham, Lopez-Bote, 1998 or Santos e Silva et al., 1999). Furthermore, the perspective of product quality depends very much on the view of the observer. The farmer has different objectives compared to the butcher or the meat industry. Finally, the consumer expects good products derived from healthy animals, reared in an environmentally friendly manner. At the same time, price remains the most important factor at the point of sale, and a low product price is normally associated with a low product quality.

The diverse aspects of pork quality have been discussed in the session on "Quality of Meat and Fat as Affected by Genetics and Nutrition" at the 50th annual meeting of the EAAP in Zurich. This publication contains the reviews and papers presented at that meeting and gives an overview from the different perspectives of meat quality and its use in human nutrition: The role of genetics, physiology, animal nutrition, meat consumption and human health and consumer concerns were all discussed and a further 38 short communications provided the most up-to-date knowledge on the subject of pig meat quality from a European perspective.

References

Bornet F., C. Alamowitch, &. G. Slama, 1994. Short chain fatty acids and metabolic effects in humans. In: Gums and stabilisers for Food Industry, Philips. G.O., D.J. Weldlock and T.E. Williams (editors) 7th ed. Oxford University Press, 217-229

DGE (Deutsche Gesellschaft für Ernährung e.V.), 1995. Empfehlungen für die Nährstoffzufuhr. 5.Überarbeitung / 2. korrigierter Nachdruck, Umschau Verlag, Frankfurt am Main

FAO (Food and Agriculture Organization of the United Nations), 1999. Foodcrops and Shortages, global information and early warning system on food and agriculture. No 3

FAOSTAT, 1998. Food and Agriculture Organization of the United Nations Data Base. Faostat.fao.org.

Lopez-Bote, C.J., 1998. Sustained Utilization of the Iberian Pig Breed. Meat Science 49: S17-S27

GSF (Schweiz. Genossenschaft für Schlachtvieh- und Fleischversorgung), 1997. Der Fleischmarkt im Überblick. Geschäftsbericht der GSF, Bern

J. Santos e Silva, J. Ferreira-Cardoso, A. Bernardo & J.S. Pires da Costa, 1999. Conservation and development of the bísaro pig. Characterisation and zootechnical evaluation of the breed for production and genetic management. 50th Annual Meeting EAAP, Zurich, Switzerland, 85-92

Leonhardt, M., 1988. Vitamine und Spurenelemente in Fleisch. In: Mikronährstoffe – ihre Rolle in der Ernährung. Schweizerische Gesellschaft für Ernährungsforschung-Herbsttagung

Leonhardt, M. & C. Wenk, 1999. Beitrag des Konsums von Schweinefleisch zur Deckung des Nährstoffbedarfs in der Schweiz. In: Böhme H. & G. Flachowsky (Eds.) Aktuelle Aspekte bei der Erzeugung von Schweinefleisch. Landbauforschung Völkenrode – Wissenschaftliche Mitteilungen der Bundesforschungsanstalt für Landwirtschaft (FAL) - Sonderheft 193

General Contributions

What is pork quality?

H. J. Andersen

Department of Animal Product Quality, Danish Institute of Agricultural Sciences, Research Centre Foulum, P.O. Box 50, DK-8830 Tjele, Denmark

Summary

At present significant variations for most quality parameters occur between pig carcasses. Unfortunately, up to now only few in the pork industry and in the primary production of pigs have realised or cared about this problem. Only lean meat content has been in focus. However, over the past ten years dramatic changes in the international market place, caused by changing lifestyles and requirements of consumers in Europe and around the World, have started to require high standards of quality assurance regarding diversity, quality and safety of products and the environmental, ethic and animal welfare aspects of their production. Consequently, the concept "pork quality" is developing and it includes, besides composition and size, eating quality, nutritionally quality, technological quality, health and hygienic quality and ethical quality. This paper will especially focus on quality aspects expected to be of outmost importance for the customers, hereby meaning the meat processing industry and the consumers, of pork meat now and into the next millennium.

Keywords: Pork quality, Pork quality characteristics, Technological quality, Eating quality

Introduction

QUALITY – What does it mean? Quality is a philosophical made-up word introduced by *Cicero* – derived from the Latin word *qualis,* which means "how" or "of which character". This implies that quality can be translated to "how like" or characteristics.

The modern use of "quality" raises problems – how to measure and evaluate quality, what is measured and, finally, which scale has to be used? Without a fairly precise idea of what is measured and how the measurement is performed, a term as "quality assurance" makes no sense.

With this in mind it cannot wonder that quality has different meanings to different people. This indeed is also the case with pork quality. For the pig producers, pork quality equals those properties, which rise the most favourable price when selling the pig to the slaughterhouse. In Europe this mainly means lean meat and sometimes a well-proportioned carcass. Consequently, the pig producers only raise pigs, which give lean meat with minimum production costs (pigs with optimal performance). This explains why PSE pork from pigs carrying the halothane gene developed so extensively for many years. Moreover, we have to be aware that the potential of introducing alike "sick" genotypes still exists, as long as no further demands are required in the main production of European pigs. In contrast, performance of the slaughter pig will in no way be a factor in the quality considerations made by the slaughterhouse. Here, main parameters in the evaluation of the pork quality will be absence of pathogens, water holding capacity (WHC), composition of the meat, microbial load, presence/absence of residues and contaminants, together with specific physical/chemical properties of value in further sale. More or less the same parameters will be valid in relation to quality requests made by the meat processing industry. Finally, changing to the consumers of meats, completely other thoughts are considered in the evaluation of the pork quality. These mainly include sensory parameters of the heated product (tenderness, juiciness and flavour) beside especially safety aspects and appearance of the product.

From the above, it is easy to understand that the question "What is pork quality?" at this time is highly relevant, where food and food production including pork need to become more and more holistic to fulfil future market demands as stated beyond. The need for a more holistic perspective in nowadays food production, is due to the dramatic changes in the international market place during the past ten years, caused by the changing lifestyles and requirements from consumers in Europe and around the world. Consumers have started to require high standards of quality assurance regarding diversity, eating quality, and safety of products and ethical, environmental and welfare aspects of their production. This means that the food quality term, including pork quality, has turned dynamic. Moreover, it cannot solely be limited to the objective descriptions of the properties of the product, as suggested by Honikel (1993) - even though this would make things "easier".

Quality Characteristics

The customers of fresh pork are the meat processing industry and consumers who buy 65-80% and 20-35%, respectively, of the pork produced as a whole. It is important to have these numbers in mind in the discussion of pork quality. Previously, this has seldom been the case, as practically all the emphasis has been on consumer associated quality characteristics. Consequently, characteristics demanded or needed by the food processing industry should have more focus in the future, even though the importance of the quality characteristics, demanded by consumers for fresh pork, still should be weighted a bit higher than the above numbers justify, as the consumers' experience with the fresh meat to a certain degree also may reflect their purchase of certain processed pork products (yet not bacon and dry cured ham, which the consumers hardly combine with fresh pork anymore). In Table 1 are groups of the most important quality characteristics presented.

Quality Characteristics Demanded by Meat Processing Industries

Technological and hygienic quality characteristics, as mentioned in Table 1 and outlined in more detail below, will without doubt satisfy most of the demands by the meat processing industry now and in the future.

Technological quality

This term contains all the attributes of value in further processing of fresh pork including both general meat processing and other food processing (e.g. catering industries).

Water holding capacity (WHC) is the most urgent technological quality attribute. WHC includes the ability of the fresh pork to retain the water in the meat and bind extra water. The higher WHC the more valuable will the pork be for use in highly processed pork products. Initiatives to improve WHC of pork have consequently high priority in the pork industry (elimination of PSE, red soft and exudate (RSE) and acid meat). Thus pH is likewise a compelling technological quality attribute, as pH_{24h} of pork is highly correlated to the WHC of the meat. pH_{24h} around 5.8 is preferable due to acceptable processing quality (e.g. reasonable WHC and good slicebility of derived pork products) and good microbial stability.

Even though higher pH results in better WHC of pork, it also results in inferior colour, flavour and microbial stability of the product.

Lipid characteristics are another factor of importance for the technological quality of pork. If the pork lipid becomes to unsaturated, the pork is not suited for e.g. sausage production. Furthermore, products become oxidative unstable which accelerate rancidity problems. The anti-oxidative status of the meat, e.g. content of vitamin E, is becoming an important technological quality attribute in pork, which is meant for the use in the production of different kinds of convenience meat products. Most of these are pre-cooked and thereby have problems with formation of inferior flavour development upon re-heating, development of warmed-over-flavour (WOF). An extensive amount of research has shown that dietary supplementation of vitamin E to pigs is an effective tool in minimisation of WOF. Finally, eating quality attributes are slowly becoming technological quality attributes in pork used in processing industries specialising in convenient food servings.

Table 1. Groups of pork quality characteristics.

Group	Individual attributes	Remarks
Eating quality	Appearance Flavour Tenderness Juiciness	*(see text for further details)*
Nutritional quality	Protein content/composition Lipid content/composition Vitamins Minerals Digestibility	
Technological quality	Water holding capacity pH-value Protein content and its status Lipid content and its characteristics Content of connective tissue Cutting piece/size Anti-oxidative status	e.g. non-denatured vs. denatured Saturated, mono or polyunsaturated e.g. vitamin E content Tenderness *(see text for further details)*
Hygienic quality	Microorganisms Residues Contaminants	Both pathogens and spoilers e.g. antibiotics, hormones & other pharmaceuticals e.g. pesticides, heavy metal ions etc. *(see text for further details)*
Ethical quality	Organic farming Religion Outdoor rearing Welfare aspects	 e.g. pork not allowed, ritual slaughter procedure etc. e.g. no use of growth promoters

Hygienic quality

Hygienic quality includes presence, distribution and numbers/concentration of spoilers, pathogens and residues in the carcass and the different cuts. Notifications of food poisoning to doctors are increasing all over Europe and in particular, the increase in salmonella food poisoning incidents, e.g. *Salmonella thyphimurium* DT104, has been associated with pork. Use of growth promoters in pig production is speculated to give rise to emergence of resistant bacterial strains. These concerns have already led to bans on the use of certain growth promoters in EU and complete elimination of growth promoters in pig production in some of the Nordic countries.

If pigs are carriers of food poisoning organisms they are usually without symptoms. This is a problem, as classical meat inspection cannot expect to be protective. Another problem is that the mechanisms of colonisation of the pigs are often poorly understood, which is the reason why control measurements on the farm are ineffective or not available. This means that apparently healthy animals arrive at the slaughter-house and might infect other animals, and with cross contamination in the plant, bacteria can quickly be distributed. Feed is a major potential source of salmonella infections. Recent data show that levels of bacteria can be considerably reduced by instigating a programme of strict hygiene controls in feed mills including operation of the pelleting machine over 81 °C, hereby reducing incidence of salmonella (Lillie, 1995). Good manufacturing practice (GMP) at the abattoir (e.g. eliminating faecal contamination of carcasses during evisceration, low temperature in the cold chain), combined with strict hygiene control in feed production and good management (e.g. use of all in all out systems), are now accepted to be an effective way to raise the hygienic quality of pork.

Antibiotics were introduced in the 1940s in the cure of diseases and infections. At that time little thought was given to the possibility that continuing use might lead to development of resistant strains and types of pathogens in both human and pig patients. The widespread use of antibiotics for growth promotion and to improve feed utilisation efficiently in pigs has, as mentioned above, begun to raise concerns about the effects that these drugs might have once they are a continuing feature of the food chain. Consequently, legislation has been drawn up in most European countries to control the use of these agents in pig production.

Quality Characteristics demanded by the Consumers

In contrast to the quality demands by the meat processing industry, the consumer's experience of pork quality is much more complex. Consequently, it might be more convenient to divide consumer experience of pork quality into "hidden" and "visible" quality characteristics, as schematically shown in Table 2. The hidden quality characteristics include mainly safety aspects. The consumer has no way of telling whether the pork is safe or not at the time of purchase or consumption (she/he may of course be ill or even worse later!). In contrary, he/she must rely on good manufacturing practices as well as previous experience or tacitly hope that the responsible food inspector authorities are protecting him/her sufficiently.

Table 2. "Hidden" and "Visible" pork quality characteristics.

Groups	Attributes	Expectations & assumptions
"Hidden" pork quality	Safety	Absence of pathogens, toxins, contaminants & other harmful substances
	Nutritional value	Wholesome, nourishing, good protein source, functional iron source
	Image/reputation	Good, reliable
	Ethical	Organic, no use of growth promoters, outdoor rearing, good animal welfare, ritual slaughtering procedure
	Labelling	All correct
"Visible" pork quality	Appearance	Appealing
	Flavour	Expected meaty
	Tenderness	Good
	Juiciness	Good
	Convenience	Functional & sensory properties satisfactory at the time of consumption
	Price	Cheapest possible in relation to expected quality

(modified after Roschnik, 1989)

The nutritional value is also "hidden" for the average consumer, who will assume that the nutritional labelling is correct and that legal and regulatory requirements in this respect are fulfilled. With the exception of a few situations, where the fat content is given, fresh meat is seldom labelled with nutritional information.

Image and reputation of a product are critical attributes of pork. Successful business depends on the confidence the customer has in a particular "manufacturer". Overall image and reputation of fresh pork are absolutely vital, if the slaughterhouse or the manufacturer of processed pork is to remain in business. This may be exemplified in cases, which are already known to give problems in the sales of fresh pork, for example, pork has a negative image in the public (fat), despite nutritionists proving the opposite, and periodically falling market shares due to adverse publicity (e.g. Salmonella, use of growth promoters etc.).

Except for image and reputation, hidden quality characteristics have until recently had only little effect as to whether foods are purchased or not, as they have been taken for granted (Roschnik, 1989). However, as "soft" quality characteristics (animal welfare, environmental influence of production, organic & ethical production), which fall into the "hidden" category, are becoming more and more important in the consumer's choice of fresh meat, the reliability of food inspection authorities responsible for these quality aspects are becoming more and more important.

In 1995 David Lister drew attention to the ongoing change in consumer perceptions of meat quality including hitherto "unknown" quality characteristics (e.g. the above mentioned "soft" parameters), which had to be taken into serious consideration in future sales of meat, if this was to stay steady or to rise. This, he nicely stated in his synopsis of his presentation – *"Perceptions of meat quality are changing dramatically. Though they may be capricious, new considerations are not simply those which allow us to enjoy the meat eating habit but could determine whether people will continue to eat meat at all. Though this is unlikely to be of catastrophic consequence for the meat industry world-wide, elements of these will need to be incorporated into our thinking about the quality of meat in it's widest sense if it is to continue to be enjoyed and to contribute to the improved nutrition of the world's peoples"* (Lister, 1995).

Of the "visible" quality characteristics the sensory properties of pork undoubtedly provide the most important reasons for its acceptability, as most of us eat meat for pleasure. Despite individual preferences regarding sensory properties of pork are subjective - appearance, tenderness, flavour and juiciness are known to be the most important factors in the consumer acceptance of pork.

Appearance

Even though appearance does not have much to do with the eating quality of pork, this single quality parameter is becoming increasingly important in determining the consumer's purchase of pork. This is due to the fact that the majority of fresh pork is sold from displays of supermarkets and self-service retail stores, where the appearance of pork becomes the customer's single parameter to evaluate the quality of the product in the purchase situation. Consequently, consumers look for colour, fluid retaining characteristics and fat content of pork in the hope that they will indicate the eventual enjoyment of the product when it is eaten.

Flavour

Flavour of pork is the result of the formation of a large number of compounds formed during heating of the product. The flavour development mainly depends on constituents in the fresh meat, e.g. fat composition, peptides, glycogen concentration, vitamin content, especially thiamine and vitamin E etc., and the heat treatment of the product. Increasing temperatures increase flavour development. During recent years the intensity in pork flavour seems to have decreased, most probably as a result of the production of pork with a minimal content of intra-muscular fat. However, off-flavours are potentially a larger problem associated with pork palatability than the intensity or desirability of the flavour. Boar taint and lipid oxidation are among conditions associated with off-flavours in pork. Boar taint is produced by a steroid (androstenone) and a degradation product of tryptophan, skatole, which are both deposited in the fat and released upon heating. For an extensive review see Bonneau (1998). Lipid oxidation, better known as rancidity, occurs in the unsaturated lipid fraction during prolonged storage (freezer) and upon re-heating of the pork.

Tenderness

For years the tenderness of beef has been recognised to be the single most important factor in consumer acceptance (Bratzler, 1978), as this attribute is often the cause of consumer complaints (Pearson, 1994). In contrast, tenderness is a somehow neglected quality attribute in the eating quality of pork. However, there is increasing evidence emerging which shows improved palatability of the pork at the time of eating when tenderised (Taylor *et al.*, 1995).

The contractile state of the muscle (sarcomere length) after *rigor mortis* is a major factor in meat tenderness with extended sarcomere length associated with increased tenderness. Sarcomere length is in general not affected appreciably in most of the populations used for slaughter pigs, however, *post mortem* conditions might affect sarcomere length creating differences in tenderness.

Extreme cooling, chilling of the carcass below 15 °C prior to onset of *rigor mortis*, may induce the phenomenon cold-induced shortening. This results in short sarcomeres and is related to toughening in meat (Bendall, 1975). Even though this is not a huge problem in pork, pelvic suspension has been found to counteract cold-induced shortening in pigs, when rapid cooling has been performed (Møller & Vestergaard, 1986; Taylor *et al.*, 1995).

Ageing, often called conditioning, of fresh pork can be used to improve tenderness. Conditioning (ageing) of pork is a process during which a continuous weakening of the structural elements by different endogenous muscle peptidases (most probably the calpain system and cathepsins) takes place. The degree of conditioning applies to physiological attributes of the growing animal and interrelate with conditions present in the muscle in the early *post mortem* period, however, in general this leads to increased tenderness in the final product.

Juiciness

Juiciness of pork is associated with the amount of moisture present in the cooked product and the amount of intra-muscular fat. Increase in ultimate pH of the pork is associated with increased moisture retention in the heated product. In contrast, low pH early *post mortem*, which gives rise to pale, soft and exudative (PSE) meat, is closely related with low moisture content of cooked pork.

Degree of doneness has a dramatic effect on the juiciness of pork (Simmons *et al.*, 1985). Increasing internal temperature from 60 to 80 °C results in a decrease in juiciness and moisture content of heated pork.

Other Visible Quality Attributes

As seen in Table 2 convenience is also considered an important visible quality attribute. Convenience contains serving size and packaging, however, it also includes the ease of preparation (e.g. cooking time) and availability. This last factor depends on distribution system and the shelf-life.

Finally, price is an extremely visible attribute of pork related to quality by the notion "value". When purchasing any pork product, the customer assesses its "other" quality characteristics, i.e. the degree to which his/her needs and expectations have been met in relation to the price given. However, price is sometimes an odd "acceptance factor". The best way to illustrate this matter, may be through a fresh meat related example. At the time of identification of BSE in cattle in the UK, there was an immediate stop of all export, which lowered the price of beef considerably in the UK. Initially, this gave rise to a shortage in beef on sale, as demand exceeded supply in certain areas of the UK, despite uncertainty whether BSE should be considered a zoonotic disease, before the well-known massive decrease in sales of beef in the UK became the consequence.

Factors Affecting Pork Quality

Even though factors affecting pork quality are outside the scope of the present paper and will be partly covered in the subsequent papers, a few comments on this area will be given due to its high relevance in the discussion of pork quality.

On-Farm Factors Affecting Pork Meat Quality

Over recent years a considerable amount of research effort has been focused on the possibility of improving quality characteristics on the farm by manipulating animal management practices. In fact this emphasis has been provoked by customer demands. An example is the producers' success in responding to market pressures to produce leaner meat, which has led to more than a 50 % reduction in backfat thickness and a simultaneous increase in lean meat content over a period of 20 years in certain European countries. This substantial improvement in carcass lean content is the result of a combination of better genetics and improved nutritional and management practices. This is an excellent example of the pig producers' ability to respond to market pressures when the factors influencing pork quality are directly under the control of the producer.

Genetics are known to have an influence on quality attributes relevant for the customers. Some known major gene effects (e.g. the halothane and RN⁻ gene) give rise to inferior quality characteristics in fresh pork, which subsequently have huge economic consequences in the further processing of pork (decreased WHC & yield). Others, for example the recently identified IMF gene (Gerbens, 1998), may be used in the optimisation of eating quality. However, when major gene effects are excluded, genetics have been estimated to account for only 30 % of the variation found in most pork meat quality characteristics (de Vries, 1997). Therefore, other environmental factors and the poorly understood interaction between genetics and environmental factors on resulting pork quality characteristics need much more attention in the future, if many of the demands are to be accomplished.

Until recently, only potential problems with unpleasant aromas, caused by definite feed ingredients, (e.g. certain types of fishmeal), have been addressed in relation to the impact of nutrition on pork eating quality. However, the increasing interest in improving the oxidative stability of fresh pork (improved colour stability and reduced rancidity development) have during the recent years resulted in a high number of studies examining the influence of dietary vitamin E on the oxidative stability of fresh pork. These studies have shown interesting results, especially in relation to improvement of eating quality of pre-cooked pork products. Comparison of *ad lib* regimes with some degree of restriction in feeding levels have indicated that *ad lib* feeding is superior with regard to resulting eating quality characteristics (Ellis *et al.*, 1990; Warkup *et al.*, 1990). This observation may be highly relevant at a time when outdoor rearing and organic production of pigs are becoming popular in some European countries, as also supported by a recent study (Danielsen *et al.*, 1999), where restricted feeding resulted in inferior tenderness in the loin. Moreover, preliminary data have shown that strategic finishing of pigs can down-regulate muscle glycogen stores which are known to influence pork quality characteristics directly (Petersen *et al.*, 1998). Further research is required to evaluate the latter concepts for potential use as quality control tools in the production of pork meat.

Hardly any studies have been performed regarding the influence of welfare associated management factors (e.g. intensive versus free-range housing etc.) on most pork quality parameters. But, due to the continuous change in demanded quality, for example the introduction of the above mentioned "soft" quality parameters, it has become necessary to take some of these into consideration in the production of certain pork qualities. This tendency will no doubt increase in the future. The effect of such welfare associated management changes is most probably neither "visible" nor measurable, which is the reason why they need to be incorporated in quality assurance schemes (see later) to be effective.

Procedures around Slaughter and Post mortem Influencing Pork Meat Quality

As stress up to or immediately before slaughter has for many years been known to have considerable influence on quality of fresh pork, a considerable amount of research has been performed within this area. Transport time, loading procedures, mixing of unfamiliar animals before and during transport or at lairage, lairage time and stunning methods are all factors which are known to affect fresh pork quality to different extents. Time of debleeding, scalding and dehairing may also influence meat quality characteristics even though not much data is available at the present time. Furthermore, in this respect *post mortem* chilling procedure is very important, as wrong chilling procedure can lead to inferior pork quality attributes (e.g. growth of spoilers/pathogens, poor tenderness development, decrease in WHC etc.), while optimal chilling can even minimise/inhibit potential inferior development in pork quality. Finally, electrical stimulation, suspension procedure and cutting procedure are also important in respect to the final quality of pork meat.

Registration of Meat Quality Characteristics

To control variation in pork quality, use of registration methods, which can measure or predict the individual quality attributes, becomes necessary. Carcass evaluation (grading) is almost the only method used all over Europe in the prediction of "quality". However, the percentage of lean meat is primarily a quantitative, not a qualitative measure. Therefore, there is an urgent need to begin implementing methods to measure/predict pork quality, before it reaches its customers, as they will now start to expect certain quality standards to be fulfilled. Accordingly, quality assurance schemes from "farm to retail" are starting to see the daylight in the pork industry (see below). These especially include the food safety aspect with the reduction in levels of food-borne zoonoses as main target, however, schemes including animal welfare aspects are also functioning in certain European countries.

As mentioned above regarding safety, WHC and eating quality characteristics, these are critical quality attributes when the processing industry and the consumers purchase pork. Consequently, control or prediction of the development in these attributes should have first priority in the pork industry. During the years solid objective methods have been developed to determine presence of pathogens, WHC and most of the eating quality attributes. However, most of these methods cannot measure the individual attributes within 24 hour post slaughter or before the pork has been cooked. This is the main problem in measurement and prediction of pork quality, and at the present time no existing control or prediction methods are solid enough to be used in the quality control of these attributes shortly after slaughter. During the years much effort has taken place to develop such methods, but without considerable success. However, introduction of such methods is a must, if pig producers are to be paid for the quality they deliver, and if the slaughterhouse's reliability to its customers is to be improved in the future.

Quality Assurance Schemes

From the above and as discussed in more details in the subsequent paper of Jamison (2000), it is clear that the customer's definition of pork quality is now becoming wide and has been conditioned by changing attitudes in society generally, often amplified by the media. In order to strengthen the markets they presently have, the pork industry must be pro-active and build quality into pork production systems. Development of quality assurance (QA) schemes with the aim of providing assurance by defining the aspects of quality, which are most important to customers and especially consumers, are obvious for such purpose. QA schemes to satisfy customer demands regarding quality, are already common in various industries. In the pork industry, schemes are rapidly developing and are beginning to embrace many of the quality attributes that are important to customers from "gene to fork". QA schemes are a possibility to raise especially consumer confidence. Even though tightly regulated QA schemes which fulfil all future quality demands at present seem impossible to carry out, the existence of some traditional products (e.g. Parma ham) and organic meat production schemes, which operate with strict specifications, show the potential when the tools are available. For more comprehensive information on QA schemes in the meat industries the excellent paper by Wood and colleagues should be encountered (Wood et al., 1998)

Conclusions and Implications

The above-mentioned components in the total pork quality concept of today make it necessary to take elements at all stages of the pig meat chain into consideration, i.e. from the gene to the farm and through to the consumer, if quality perception and/or attributes demanded by customers now and in future are to be included in supplied pork.

The majority of the produced pork (60-70 %) has to satisfy the quality demands of the meat processing industry and other food industries, which use fresh pork in their production (e.g. convenience food, catering etc.). Consequently, the part of the pork sector that wants to supply the future "bulk" production needs to focus on strategies which ensure uniform pork quality with superior hygienic and technological quality characteristics.

The remaining part of the production, going directly to fresh meat consumption, should be considered to be part of a global niche production. Individual preferences depending on local markets (habits, upbringing, craze etc.) will determine the demands on the different markets. The part of the pork sector, which decides to supply these markets, has accordingly to focus on the demands of the specific markets. However, safety, eating and nutritional quality characteristics should have highest priority in the production, as evidence suggests that intrinsic cues (cues that are part of the physical product) usually carry more weight in formation of quality expectations by consumers than extrinsic ones (Steenkamp, 1989; Steenkamp & van Trijp, 1996). Secondly, other elements like traceability, labelling or marketing should be integrated in the total quality concept. In between or as a final element lie ethical quality attributes depending on local market attitudes towards ethical aspects in pig production.

The majority of European pork production can at present be considered as bulk production even though the "quality" is different in different countries. Accordingly, that part of the European pork sector that wants to be competitive in the coming customer driven markets, needs to take part in a difficult reorientation process. Preliminary attempts are taking place in the form of introduction of different QA schemes in certain countries, however, many of the tools, which are necessary to achieve a continuous optimal deliverance of high quality pork fulfilling customer demands, are not available at present. Therefore, immediate initiatives to develop and implement the necessary tools should be carried out as quickly as possible. The two most obvious hurdles that have to be overcome before such a process can proceed are;

- *Development of solid on-line/at-line methods that can measure demanded quality attributes at the slaughterhouse. This will make it possible to pay the producers for the quality they deliver and supply customers with given qualities.*

- *Convince animal scientists occupied with pig production that product quality is a function of productivity and not the other way around, as assumed ever since establishment of this discipline.*

If these two hurdles are overcome, it will, without doubt, be possible to propose strategies for production of pork of high quality, including attributes demanded by customers, in the foreseeable future. These should subsequently be introduced in the pig meat chain in combination with present knowledge of slaughter technologies that preserve or even improve the quality of the pork during its transformation from muscle to meat.

Overall it is a major challenge for animal -, meat – and food engineering science to unite their potential in the time to come and subsequently show that it is possible to set up a continuous production of pork of a quality demanded by the customers. In return the dynamic nature of pork quality will ensure a continuous need for an integrated effort from animal -, meat - and food engineering science for all time.

References

Anderson, K.R., J.S. Petersen, H. Johansen, S.K. Jensen, A. Karlsson & H.J. Andersen, 1998. Feed-induced muscle glycogen changes in slaughter pigs and their influence on meat quality. "Proc. 44nd International Congress of Meat Science & Technology" Vol. 1, Barcelona, Spain, 276-277

Bendall, J.R., 1975. Cold contracture and ATP-turn-over in the red and white musculature of the pig, post-mortem. J. Sci. Food Agricul. 26: 55-71

Bonneau, M., 1998. Use of entire males for pig meat in the European Union. In: Proceedings 44th International Congress of Meat Science and Technology Vol. I, Barcelona, Spain, 192-205

Bratzler, L.J., 1978. Palatability characteristics of meat. In: Science of Meat and Meat Products, J.F. Price & B.S Schweigeri (editors), Food and Nutrition Press, Westport, Connecticut, USA, 328-363

Danielsen, V., L.L. Hansen, F. Møller & C. Bejrholm, 1999. Produktionsresultater og spisekvalitet i svinekød fra grise fodret med forskellig mængde grov- og kraftfoder. In: Internal report no. 117, Grovfoder og fiberrige fodermidler til svin, K. Jakobsen & V. Danielsen (editors), Danish Institute of Agricultural Sciences, 47-50. (Danish)

de Vries, A.G., 1997. PIC Europe (personal communication).

Ellis, M., A.J. Webb, P. Avery, R. Smithard & I. Brown, 1990. Sex and feeding effects on the eating quality of fresh pork. Animal Production 50: 551-555

Gerbens, F., A. Jansen, A.J.M. Van Erp, F. Hardens, T.H.E. Meuwissen, G. Rettenberger, J.H. Veerkamp & M.F.W. Te Pas, 1998. The adipocyte fatty acid binding protein locus: Characterization and association with intramuscular fat content in pigs, Mamm. Genome 9: 1022-1026

Honikel, K.O.1993. Quality of Fresh Pork – Review. In: Pork Quality: Genetic and Metabolic Factors, E. Puolanne, D.I. Demeyer, M. Ruusunen & S. Ellis (editors), CAB International, UK.203-216

Jamison, W., 2000. Who eats meat: factors effecting pork consumption in Europe and the United States. In: Genetic and Pork Quality 111-118

Lillie, A., 1995. Preventing zoonotic agents in domestic animals by proper feed hygiene. In: Meat Quality and Safety as Affected by Primary Production, A.J. Møller, M.M. Mielche & P.A. Barton-Gade (editors), Eur. Cons. Cont. Edu. Adv. Meat Sci. Technol., Utrecht, The Netherlands, 195-201

Lister, D., 1995.The meat we eat: Notions of quality for today and tomorrow. In: Proceedings 41st International Congress of Meat Science and Technology Vol. II. San Antonio, Texas, USA, 3-12

Møller, A.J. & T. Vestergaard, 1986. Effects of altered carcass suspension during rigor mortis on tenderness of pork loin. Meat Science 18: 77-87

Pearson, A.M., 1994. Introduction to quality attributes and their measurement in meat, poultry and fish products. In: Quality attributes and their measurement in meat, poultry and fish products, A.M. Pearson & T.R. Dutson (editors), Blackie Acad. & Professional, London, UK, 1-33

Roschnik, R., 1989. New trends in consumer acceptance and convenience. In: Proceedings of Euro Food Chem V, Versailles, France, 30-39

Simmons, S.L., T.R. Carr & F.K. McKeith, 1985. Effect of internal temperature and thickness on palatability of pork loin chops. J. Food Science 50: 313-315

Steenkamp, J.-B., 1989. Product Quality: An investigation into the concept and how it is perceived by consumers. Assen: Van Gorcum

Steenkamp, J.-B.& H. van Trijp, 1996. Quality guidance: A consumer-based approach to food quality improvement using partial least squares. European Review of Agricultural Economics 23: 195-215

Taylor, A.A., G.R. Nute & C.C. Warkup, 1995. The effect of chilling, electrical stimulation and conditioning on pork quality. Meat Science 39: 339-347

Taylor, A.A., A.M. Perry & C.C. Warkup, 1995. Improving pork quality by electrical stimulation or pelvic suspension of carcasses. Meat Science 39(3): 327-337

Warkup, C.C., A.W. Dilworth, A.J. Kempster & J.D. Wood, 1990. The effect of sire type, company source, feeding regimen and sex on eating quality of pig meat. Animal production 50: 543-550

Warner, R.D., R.G. Kaufman & M.L. Greaser, 1997. Muscle protein changes post mortem in relation to pork quality traits. Meat Science, 45: 339-352

Wood, J.D., J.S. Holder & D.C.J. Main, 1998. Quality assurance schemes. In: Proceedings 44th International Congress of Meat Science and Technology Vol. I, Barcelona, Spain, 206-215

Influence of genetics on pork quality

A.G. de Vries*[1], L. Faucitano[1], A. Sosnicki[3] & G.S. Plastow[2]

[1]*PIC Europe and* [2]*PIC Group, Fyfield Wick, Abingdon, OX13 5NA, UK;*
[3]*PIC Americas, P.O. Box 348, Franklin, KY 42135-0348, USA*

Summary

The aim of this paper is to discuss the opportunities of gene technology in relation to the quality of pork. After dealing with breed effects and within-breed variation, an overview of major genes and DNA technology is given. It is demonstrated that some of the breed effects can be fully explained from the presence of a single gene with major effect. Within breeds, there is considerable genetic variation in relevant meat quality traits like waterholding capacity and intramuscular fat. Again, part of this variation is due to major genes. As a result, DNA marker technology can play an important role in improving meat quality. Selective breeding based on this technology will also increase the uniformity of the final product. Furthermore, the exploitation of major genes can be highly relevant for differentiation of breeding populations for specific markets.

Keywords: Pigs, Breeding, Meat quality, Major gene, DNA markers

Introduction

The quality of raw pig meat is influenced by a large number of genetic and non-genetic factors. The latter include farm, transport, slaughter and processing conditions. Meat scientists have performed a substantial amount of research on these factors, which has led to considerable quality improvement. Part of the research has also been dedicated to the genetic background of the pigs, and several studies have revealed the importance of genetic factors (e.g. Sellier & Monin, 1994). This has made the industry aware that selective breeding of pigs and the use of gene technology can play an important role in enhancing pork quality (Tarrant, 1998).

The aim of this paper is to discuss these genetic opportunities. After discussing differences between breeds and within breeds, we will focus on exploitation of individual genes based on DNA technology.

Breed Effects

Significant breed effects have been reported for intramuscular fat, waterbinding capacity, colour and tenderness (Sellier & Monin, 1994). Piétrain and Belgian Landrace pigs can give meat with inferior quality when compared to Large White or French Landrace (Monin *et al.*, 1986; Touraille & Monin, 1994). Due to a fast pH decline after slaughter, their meat can be pale, exudative and less tender. However, this breed effect can probably be completely explained through a high frequency of one single gene, the so-called Halothane gene (see later under gene effects).

Meat from Hampshire pigs often shows a markedly lower ultimate pH. This leads to a lower waterbinding capacity and higher cooking loss (Monin & Sellier, 1985). On the other hand, this breed is often characterised by higher tenderness (Sellier & Monin, 1994). Again, the specific effect of Hampshire has been found to be related to a single gene (RN), which will be discussed later.

The Large White and Duroc breeds are seen as positive contributors to meat quality. Also the meat from Landrace can be of high quality, provided that the Halothane gene has been eliminated. An extra benefit of the Duroc in some markets is the often two-fold higher % of intramuscular fat when compared to Large White and Landrace pigs (Lo *et al.,* 1992; Armero *et al.,* 1998), which can positively contribute to eating quality. Indeed, as the percentage of Duroc genes increases (0, 25%, 50% and 75%) meat is scored by panellists as being more juicy, more tender and with higher pork flavour and lower abnormal flavours (MLC, 1992). Certain studies also report a higher haem pigment content for Duroc (ao. Garrido *et al.,* 1998).

Several comparisons of meat quality between European and American breeds with Chinese purebred or crossbred pigs revealed that the latter provide more tender, more juicy and tasty product (Touraille *et al.,* 1989; Suzuki *et al.,* 1991). However, the amount of visible fat was judged as excessive in meat from Chinese crossbreeds, which counterbalanced the better meat quality. In addition, Ellis *et al.* (1995) reported no evidence of any beneficial effect of the Chinese Meishan breed on overall meat quality.

Traditional European breeds have been studied in more detail in recent years. Corsican pigs were found to have a better waterbinding capacity than Large White (Goutefongea *et al.,* 1983; Casabianca & Luciani, 1989). Iberian pig breeds were studied by Sañudo & Sierra (1989) and Serra *et al.* (1998). In a comparison with Landrace, the latter study found higher levels for ultimate pH, haem pigment content, intramuscular fat, type I fibres and lower concentrations of polyunsaturated fat. This was accompanied by large differences in age at slaughter and backfat levels. Comparative studies which involved traditional British breeds (Tamworth, Glouchester Spots, Saddleback and others) and improved breeds (LW, P, L, H and D) showed that the latter ones had paler and more watery meat. The meat from Tamworth scored the highest overall sensorial acceptability (Warriss *et al.,* 1996).

Within Breed

Within a breed, there is still variation in meat quality between individual pigs. Part of this variation is of genetic origin. Geneticists quantify this proportion by looking at differences between and within families, and refer to this as the degree of heritability. The first heritabilities (h^2) for pig meat quality traits were reported in the early sixties (Duniec *et al.,* 1961), and since that time numerous h^2 estimates have been published for all important meat quality attributes, including eating quality (for review see Cameron, 1990; Wood, 1990; Hovenier *et al.,* 1992, 1993; De Vries *et al.,* 1994) as well as fibre characteristics (Larzul *et al.,* 1997a). According to Wood (1990), there is moderate genetic variation for some traits (i.e. drip loss, intramuscular fat content and colour), but little opportunity exists to improve eating quality attributes such as flavour, texture or juiciness.

Selection for improved meat quality within lines has been limited by the absence of relevant measures that can be taken on live pigs. However, the existence of major genes together with the development of DNA technology may overcome this problem.

Major Genes

Geneticists consider that a gene can be defined as a major gene, when the difference between the mean value of the individuals homozygous for this gene and that of individuals not carrying this gene, is equal or superior to one phenotypic standard deviation of the trait of interest. Genes with such large effects can usually be detected by analysing phenotypic data across families where the gene segregates. This approach is referred to as segregation analysis. Sellier & Monin (1994) reviewed two important major genes that affect pig meat quality: Halothane and RN. After a short update on these two genes, two more recent examples for pigs will be provided.

Halothane Sensitivity Gene

This gene has been studied and discussed extensively. A thorough review is given by Sellier & Monin (1994), and results of Larzul *et al.* (1997b) and a review of Hermesch (1997) add interesting material to the discussion. The gene started to become relevant for breeders when Christian (1972) speculated the existence of monogenic variation in stress-susceptibility and when Eikelenboom & Minkema (1974) showed that the stress syndrome could be triggered by halothane gas. Since then, many comparisons regarding meat quality have been done, and most studies showed major differences in pH, colour and waterbinding between stress positive and negative pigs. These differences were directly related to a large difference in PSE (Pale Soft Exudative meat) incidence between the two genotypes, which can be explained by the different muscle metabolic profile. Indeed, fibre areas are shown to be larger and capillarisation (cap/mm^2) lower in stress-positive pigs, which lead to higher lactate formation in the muscle (Essen-Gustavsson *et al.*, 1992; Klont, 1993).

Since 1991 we have been able to accurately separate all three Halothane genotypes (instead of just reactors (nn) from non-reactors (NN and Nn)) with the Hal1843[TM1] DNA test. This test was developed from the work of Fujii *et al.* (1991) who found the causative mutation for porcine stress syndrome in the gene encoding the ryanodine receptor or calcium release channel (CRC1). Once it was possible to easily detect heterozygous animals, more detailed work on the effect of this mutation was possible.

RN⁻ gene

The RN⁻ gene was first suggested by Naveau (1986) and later confirmed by segregation analysis in two French composite lines of pigs (Le Roy *et al.,* 1990) as a gene with dominant inheritance. The recessive allele leads to a decreased technological quality due to a lower meat protein content and reduced ultimate pH. The latter is a result of an increased glycogen content in the white (fast-glycolytic) fibres. For (low phosphate) cooked ham processing, the yield of RN⁻ carriers is 5-6% lower, and moreover these hams can have extremely high slicing losses. So far, the gene has been found segregating only in populations with Hampshire influence, but because of its dominant inheritance and the widespread use of Hampshire derived sires, the gene is very relevant for breeders.

Intramuscular fat

Intramuscular fat (IMF) has been identified as playing a role in eating quality of pork. Janss *et al.* (1994; 1997) performed a segregation analysis on meat quality data of F2 crosses between Meishan and Dutch pig strains. They detected a recessive major gene for IMF, originating from Meishan. Animals with two copies of the gene had an average of 3.9% IMF in the loin, whereas carriers and homozygous negative animals had 1.8%. Research is underway to look at the existence of this gene in purebred populations, for example Duroc (Monin *et al.,* 1998).

---

[1] The HAL-1843[TM] is licensed from the Innovations Foundation, Toronto, Canada, owner of the trademark

This could eventually lead to DNA tests that allow better control of the marbling level of pork. Breeders will then have a large influence on the level of this trait, since its heritability is around 50% (Cameron, 1990; Hovenier *et al.*, 1993; De Vries *et al.*, 1994). The challenge is to achieve a higher IMF without increasing the levels of the other fat deposits (subcutaneous, abdominal and intermuscular). More attention to this is given in the Candidate Gene section of this paper (H-FABP gene).

Androstenone

Another trait with a high heritability is the level of androstenone (Willeke, 1993), which is one of the causes of the so-called 'boar taint' problem in meat from entire males (Bonneau, 1998). Applying segregation analysis, Fouilloux *et al.* (1997) found a major gene for androstenone level in LW populations that were selected on this trait. The gene giving rise to a low androstenone level was dominant, and carriers of this gene had 3 standard deviation (SD) units lower level than non-carriers (0.33 vs. 0.90 ppm). In the same data set, the authors also found a major gene for the development of the bulbo-urethral glands. The size of these glands are seen as a good indicator of the sexual maturity status of boars (Fouilloux *et al.*, 1997).

Earlier work in France (Bidanel *et al.*, 1996) showed a large gene effect on androstenone level in an F2 generation of Meishan with Large White. This gene effect was linked with the major histocompatibility complex of the pig (SLA). Linkage with SLA haplotypes was also shown with male genital tract development in work by Rothschild *et al.* (1986).

Muscle fibre traits

Muscle fibre types differ phenotypically in that they express different subsets of myofibrillar isoform genes with different ATPase activities as well as different types and levels of metabolic enzymes. The different myosin heavy chain isoforms are coded for by separate genes, some of which are preferentially expressed in fast skeletal muscle and in slow skeletal muscle (see review by Goldspink, 1996). The number of myofibers is prenatally determined, so the maximal number of myofibers available for meat production is formed during embryonic myogenesis, which seems to be under the genetic control of the MyoD gene family (Buckingham *et al.*, 1992). The reader is referred to the Te Pas *et al.* (1994) review for more detailed information on the MyoD gene family muscle regulatory mechanism.

DNA Technology

The evidence for major genes reported in the previous section was originally obtained using segregation analysis, i.e. without any DNA marker information. Afterwards molecular studies were performed to detect the location of these genes on the genetic map. In practice, and except for alleles of very large effect, DNA studies are required to dissect the genetic nature of most traits of economic importance. In this section an overview of the latest results is given. The first part deals with DNA markers that are closely linked with major genes. The second part deals with mutations in targeted functional genes referred to as the candidate gene approach.

DNA Markers

Markers can be used to localise genes responsible for qualitative traits like coat colour (e.g. Johansson Moller *et al.*, 1996), and they can also be used to detect genes with substantial effects on quantitative traits like growth rate, IMF etc. In this case the approach is referred to as QTL (Quantitative Trait Locus) mapping.

In pigs, the QTL mapping approach normally uses families created from crosses based on divergent lines, for example Large White with wild boar or Chinese Meishan. Indeed, this approach has been used in relation to intramuscular fat (IMF). Having identified a major gene for IMF using segregation analysis (Janss *et al.*, 1994, 1997), this group went on to QTL mapping, using microsatellite markers. They recently reported evidence for suggestive linkages with markers on chromosome 1 and a marker on chromosome 3 (De Koning *et al.*, 1998). The same region on chromosome 1 was also identified as containing a QTL for backfat. The latest analysis from this group has also identified QTLs for this trait on chromosomes 4 and 7 (M. Groenen, pers. comm.). In a recent study on Iberian x Landrace progenies a QTL for intramuscular fat was found on chromosome 6 (Óvilo *et al.*, 1999).

A number of pig populations are now being used or have been created for the purposes of searching for meat quality QTLs. These include crosses between the following breeds: Wild Boar, Large White, Meishan, Landrace, Piétrain, Iberian pig, Duroc, Mangalitsa and Berkshire (De Vries *et al.*, 1998).

Candidate Genes

The candidate gene approach can be relatively straightforward compared to the QTL approach. For example, we have used polymorphisms in candidate genes to look for associations across populations. When associations are identified the resulting marker can potentially be used directly in breeding programmes. This approach has been used very successfully for ESR and litter size (Short *et al.*, 1997).

An example for a candidate gene for meat quality is provided by the gene for heart fatty acid binding protein (H-FABP). Gerbens *et al.*, (1997) identified polymorphisms in this gene and found these to be associated with variation in IMF in the Duroc (Gerbens, 1998a). H-FABP maps to pig chromosome 6 and not to the QTL regions identified by De Koning *et al.*, (1998), (see the section on Major Genes). A comparison of the homozygous haploid classes found that they differed by about 15% of the mean value. More recently, this group has found a larger effect on IMF with the related gene adipocyte FABP (FABP4. Gerbens *et al.*, 1998b). In addition, the effect appears to be independent of backfat and so offers promise for the manipulation of IMF by MAS. These candidate gene markers have now been included in a joint analysis with the QTL analysis of Janss an co-workers and it will be interesting to see if the variation in IMF in the crosses can now be explained more completely (see above).

Discussion

From breeds to genes

The early work on genetic effects on meat quality focused on breed differences, and those differences are still very relevant. However, a number of specific breed effects have more recently been found to be caused by single major genes (e.g. Halothane in Piétrain, RN gene in Hampshire). This has encouraged the approach of examining single gene effects, rather than breed effects. It has also allowed the breeding organisations to make use of a wider range of breeds, as it is possible to change the frequency of favourable or unfavourable genes through selective breeding. With a combination of backcrossing and selection, it is even possible to move a foreign favourable gene into a breed. The latter procedure is referred to as gene introgression.

Exploitation of major genes for meat quality

Improving meat quality is not just about changing levels of traits like tenderness or marbling, but it is also about increasing uniformity. The existence of major genes provides excellent opportunities for improving meat quality, since it allows large steps to be made in the desired direction (e.g. improving technological yield of ham process by selecting against RN⁻ gene in pigs). Secondly, it will help to reduce variation, since we can fix relevant genes in our products. Another aspect is that major genes allow differentiation for specific markets. For example, in certain types of dry cured ham a high IMF is required, whereas other products like cooked ham require a low amount of IMF. For the future it is expected that processors and retailers will specify a whole series of genes that have to be present or absent for each product that they process or sell.

Conclusions

- There are clear breed effects on meat quality, which in some cases are fully related to the presence of a single gene with major effect.
- Within breeds, there is considerable genetic variation in important meat quality traits, which again is partly caused by major genes.
- DNA technology provides excellent opportunities to improve meat quality in selection schemes within lines.
- Selection on major genes will not only increase average levels of quality but also decrease variability (i.e. increase uniformity). On the other hand, major genes can be exploited for differentiation for specific markets.

References

Armero, E., M. Flores, J-A Barbosa, F. Toldra, & M. Pla, 1998. Effects of terminal pig sire types and sex: On carcass traits, meat quality and sensory analysis of dry-cured ham. Proc 44th Int. Cong. Meat Sci. Techn., Barcelona, 2: 904-905

Bejerholm, C. & P.A. Barton-Gade, 1986. Effect of intramuscular fat level on eating quality in pig meat. Proc. 32nd Europ. Meet. Meat Res. Workers, August 24-29, Ghent, 389-392

Bidanel, J.P., D. Milan, C. Chevalet, N. Woloszyn, J.C. Caritez, J. Gruand, P. Le Roy, J. Lillehammer, Gellin & L. Ollivier, 1996. Chromosome 7 mapping of a quantitative trait locus for fat androstenone level in Meishan x Large White F2 entire male pigs. Book of abstracts 47th. Ann. Meet. EAAP, Wageningen Pers, Wageningen, 2: 610

Bonneau, M.,1998. Use of entire males for pig meat in the European Union. Meat Sci., 49: 257-272

Buckingham, M.,1992. Making muscle in mammals. TIG 8, 144

Cameron, N.D., 1990. Genetic and phenotypic parameters for carcass traits, meat and eating quality in pigs. Livest. Prod. Sci., 26: 119-135

Cassabianca, F. & A. Luciani, 1989. Caractéristiques de la viande de porc Corse issu d'elevage extensif. 1. Qualités technologiques de la viande. Colloque Prod. Porcine en Europe Médit., Ajaccio.

Christian, L.L., 1972. A review of the role of genetics in animal stress susceptibility and meat quality. In "Proc. Pork Quality Symp.", Cassens, R.G., F. Giesler & Q. Kolb (editors), Univ. Wisconsin, 91-115

De Koning, D.J., L.L.J. Janss, J.A.M Van Arendonk, P.A.M Van Oers & M.A.M. Groenen, 1998. Mapping major genes affecting meat quality in Meishan crossbreds using standard linkage software. Proc. 6th World Cong. Genet. Appl. Livest. Prod., 26: 410-413

De Vries, A.G., P.G. Van der Wal, T. Long, G. Eikelenboom & J.W.M. Merks, 1994. Genetic parameters of pork quality and production traits in Yorkshire populations. Livest. Prod. Sci., 40: 277-289

De Vries, A.G., A. Sosnicki, J.P. Garnier & G.S. Plastow, 1998. The role of major genes and DNA technology in selection for meat quality in pigs. Meat Sci., 49 (suppl. 1): 245-255

Duniec, H., J,. Kielanovski & Z. Osinska, 1961. Heritability of chemical fat content in the loin muscle of baconers. Anim. Prod., 3: 195-198

Eikelenboom, G. & D. Minkema, 1974. Prediction of pale, soft, exudative muscle with a non-lethal test for the halothane-induced porcine malignant hyperthermia syndrome. Tijdschr. Diergeneeskunde, 99: 421-426

Ellis, M., C. Lympany, C.S. Haley, I. Brown & C.C. Warkup, C.C., 1995. The eating quality of pork from Meishan and Large White pigs and their reciprocal crosses. Anim. Sci., 60: 125

Essen-Gustavsson, B., K. Karlstrom & K. Lundstrom, 1992. Muscle fibre characteristics and metabolic response at slaughter in pigs of different halothane genotypes and their relation to meat quality. Meat Sci., 31: 1

Fouilloux, M.N., P. Le Roy, J. Gruand, C. Renard, P. Sellier & M. Bonneau, 1997. Support for single major genes influencing fat androstenone level and development of bulbo-urethral glands in young boars. Genet. Sel. Evol. 29: 357-366

Fujii, J., K. Otsu, F. Zorzato, S. De Leon, V.K. Khanna, J.E. Weiler, P.J. O'Brien & D.H. Maclennan, 1991. Identification of a mutation in porcine ryanodine receptor associated with malignant hyperthermia. Science, 253: 448-451

Garrido, M.D., M.V. Granados, D. Álvarez, S. Bañon, J.M. Cayuela & J. Laencina, 1998. Pork quality of pig crossbreeds Large White, Landrace, Hampshire, Pietrain and Duroc. Proc 44th Int. Cong. Meat Sci. Techn., Barcelona, 1: 274-275

Gerbens, F., G. Rettenberger, J.A. Lenstra, J.H. Veerkamp & M.F.W. Te Pas, 1997. Characterization, chromosomal localization, and genetic variation of the porcine heart fatty acid-binding protein gene. Mamm. Genome, 8: 328-332

Gerbens, F., A.J.M. Van Erp, T.H.E. Meuwissen, J.H. Veerkamp & M.F.W. Te Pas, M.F.W., 1998a. Heart Fatty-Acid Binding Protein Gene Variants are Associated with Intramuscular Fat Content and Back Fat Thickness in Pigs. Proc. 6th World Cong. Genet. Appl. Livest. Prod., 26:187-190

Gerbens, F., A. Jansen, A.J.M. Van Erp, F. Harders, T.H.E. Meuwissen, J.H. Rettenberger, Veerkamp, M.F.W. Te Pas, 1998b. The adipocyte fatty acid-binding protein locus: characterization and association with intramuscular fat content in pigs. Mamm. Genome 9: 1022-1026

Goldspink, G., 1996) Muscle growth and muscle function: a molecular biological approach. Res. Vet. Sci., 60: 193

Goutefongea, R., J.P. Girard, J.L. Labadie, M. Renerre & C. Touraille, 1983. Utilisation d'aliments grossiers pour la production de porcs lourds. Interaction entre type génétique, sexe et mode de conduite. 2. Qualité de la viande et aptitude á la transformation. Journ. Rech. Porc. en France, 15: 193-200

Hermesch, 1997. Genetic influences on pork quality. In "Manipulating Pig Production" vol. VI, Eds. P.D. Cranwell, Austr. Pig Science Assoc., 82.

Hovenier, R., E. Kanis, Th. Van Asseldonk & N.G. Westerink, 1992. Genetic parameters of pig meat quality traits in a halothane-negative population. Livest. Prod. Sci., 32: 309-321

Hovenier, R., E. Kanis, Th. Van Asseldonk & N.G. Westerink, 1993. Breeding for pig meat quality in halothane negative populations – A review. Pig News Inform. 14: 17N-25N

Janss, L.L.G., J.A.M. Van Arendonk & E.W. Brascamp, 1994. Identification of a single gene affecting intramuscular fat in crossbreds using Gibbs sampling. Proc. 5th World Cong. Genet. Appl. Livest. Prod., 18: 361-364

Janss, L.L.G., J.A.M. Van Arendonk & E.W. Brascamp, 1997. Bayesian statistical analyses for presence of single genes affecting meat quality traits in a crossed pig population. Genetics, 145: 395-408

Johansson Moller, M., R. Chaudhary, E. Hellmén, B. Höyheim, B. Chowdhary & L. Andersson, 1996. Pigs with the dominant white coat colour phenotype carry a duplicate of the KIT gene encoding the mast/stem cell growth receptor. Mamm. Genome, 7: 822-830

Klont, R., E. Lambooy & J.G. Logtestjin, 1993. Effect of preslaughter anaesthesia on muscle metabolism and meat quality of pigs of different halothane genotypes. J. Anim. Sci., 71: 1477

Larzul, C., L. Lefaucheur, P. Ecolan, J. Gogue, A. Talmant, P. Sellier & G. Monin, 1997a. Phenotypic and genetic parameters for longissimus muscle fiber characteristics in relation to growth, carcass and meat quality traits in Large White pigs. J. Anim. Sci., 75: 3126-3137

Larzul, C., P. Le Roy, R. Gueblez, A. Talmant, J. Gogué, P. Sellier & G. Monin, 1997b. Effect of halothane genotype (NN, Nn, nn) on growth, carcass and meat quality traits of pigs slaughtered at 95 kg or 125 kg live weight. J. Anim. Breed. Genet.: 114, 309-320

Le Roy, P., J. Naveau, J.M. Elsen & P. Sellier, 1990. Evidence for a new major gene influencing meat quality in pigs. Genet. Res. (Camb.), 55: 33-40

Lo, L.L., D.G. McLaren, F.K. McKeith, R.L. Fernando & J. Novakofski, 1992. Genetic analyses of growth, real-time ultrasound, carcass, and pork quality traits in Duroc and Landrace pigs: I. Breed effects. J. Anim. Sci., 70: 2373

MLC, 1992. Meat and Livestock Commission, Second Stotfold Pig Development Unit Trial Results, MLC, Milton Keynes

Monin, G. & P. Sellier, 1985. Pork of low technological quality with a normal rate of muscle pH fall in the immediate post-mortem period: the case of the Hampshire breed. Meat Sci., 13: 49-63

Monin, G., A. Talmant, D. Laborde, M. Zabari & P. Sellier, 1986. Compositional and enzymatic characteristics of the Longissimus dorsi muscle from Large White, halothane-positive and halothane-negative Piétrain and Hampshire pigs. Meat Sci., 16: 307-316

Monin, G., P. Sellier & M. Bonneau, 1998. Trente ans d'évolution de la notion de qualité de la carcasse et de la viande de porc. Journ. Rech. Porc. France, 30: 13-27

Naveau, J. 1986. Contribution a l'étude du determinisme genetique de la qualité de la viande porcine. Heritabilité de Rendement technologique Napole. Journ. Rech. Porc. France, 18: 265-276

Óvilo, C., M. Pérez-Enciso, C. Barragán, A. Plop, M. Toro, M.C. Rodriguez, M. Oliver, C. Barboni & J.L. Noguera, 1999. Busqueda de QTLS en el cromosoma 6 para grasa intramuscular y pigmentos en un cruze F2 iberico x Landrace. Proc. 8th Jornada sobre Producción Animal (in press)

Rothschild, M.F., C. Renard, P. Sellier, M. Bonneau & M. Vaiman, 1986. Swine lymphocyte antigen (SLA) effects on male genital tract development and androstenone level. Proc. 3rd World Cong. Genet. Appl. Livest. Prod., 11,: 197-202.

Sanudo, C. & I. Sierra, 1989. Cruziamentos y produccion intensiva in cerdo iberico. 3. Calidad de la carne. Colloque Prod. Porcine en Europe Médit., Ajaccio

Sellier, P. & G. Monin, 1994. Genetics of pig meat quality: a review. J. Muscle Foods, 5: 187-219

Serra, X., F. Gil, M. Pérez-Enciso, J.M. Vàzquez, M. Gispert, I. Diaz, F. Moreno, R. Latorre, & J.L. Noguera, 1998. A comparison of carcass, meat quality and histochemical characteristics of Iberian (Guadyerbas line) and Landrace pigs. Livestock Production Science, 56: 215-223

Short, T.H., M.F. Rothschild, O.L. Southwood, D.G. McLaren, A. De Vries, H. Van der Steen, G.R. Eckardt, C.K. Tuggle, J. Helm, D.A. Vaske, A.J. Mileham & G.S. Plastow, 1997. Effect of the Estrogen receptor locus on reproduction and production traits in four commercial pig lines. J. Anim. Sci., 75: 3138-3142

Suzuki, A., N. Kojima, Y. Ikeuchi, S. Ikarashi, N. Moriyama, T. Ishizuka & Tokushige, 1991. Carcass composition and meat quality of Chinese purebred and European x Chinese crossbred pigs. Meat Sci., 29: 31-41

Tarrant, P.V., 1998. Some recent advances and future priorities in research for the meat industry. Proc 44th Int. Cong. Meat Sci. Techn., Barcelona, 1: 2-13

Te Pas, M.F.W., A.G. De Vries & A.H. Visscher, A.H., 1994. The MyoD family and meat production – a review. Proc. 40th ICoMST, The Hague, NL, S-VII. 08

Touraille, C. & G. Monin, 1984. Comparaison des qualités organoleptiques de la viande de porcs de trois races: Large White, Landrace français, Landrace belge. Journ. Rech. Porcine en France, 16: 75-80

Touraille, C., G. Monin & C. Legault, 1989. Eating quality of meat from European x Chinese crossbred pigs. Meat Sci., 25: 177-186

Warriss, P.D., S.C. Kestin, S.N. Brown & G.R. Nute, 1996. The quality of pork from traditional pig breeds. Meat Focus Int., 5/6, 179

Willeke, H.,1993. Possibilities of breeding for low 5α-androstenone content in pigs. Pig News Info, 14, 31N-33N

Wood, J.D., 1990. Consequences for meat quality of reducing carcass fatness. In: "Reducing Fat in Meat Animals". (Ed. Wood and Fisher) Elsevier Sci. Publishers Ltd., 344-389

Genetic parameters of meat quality traits recorded on Large White and French Landrace station-tested pigs in France

T. Tribout & J.P. Bidanel

INRA, Station de Génétique Quantitative et Appliquée, 78352 Jouy-en-Josas, France

Summary

Genetic parameters of traits recorded on slaughtered animals tested in French central test stations were estimated for the Large White (LW) and French Landrace (FL) breeds using a REML procedure applied to a multiple trait animal model. The data consisted of, respectively, 11500 and 4963 records in the LW and the FL breeds, collected from 1988 to 1999. Four performance traits - average daily gain (ADG), food conversion ratio (FCR), dressing percentage (DP), carcass lean content (CLC) - and 8 meat quality traits - ultimate pH of *Adductor femoris* (PHAD) and *Semimembranosus* (PHSM) muscles, two reflectance measurements (REFL and L*), water holding capacity (WHC), a visual score (SCOR), two meat quality indexes (MQI1, MQI2) - have been considered. Heritabilities of meat quality traits were low to moderate. Large positive genetic correlations were obtained between the two pH and the two reflectance measurements. Reflectance, pH, SCOR and the two MQI had large favourable genetic correlations. All meat quality traits, except WHC, had unfavourable genetic correlations with CLC and FCR. Conversely, meat quality traits were genetically almost independent from ADG and DP.

Keywords: Pigs, Meat quality, Genetic parameters, Heritability, Correlation

Introduction

Accurate estimates of genetic parameters of economically important traits are necessary to estimate breeding values and to optimise breeding schemes. Genetic parameters of traits recorded in French central test stations have been recently estimated (Garreau *et al.*, 1998). This study concerned growth traits, carcass traits, and a meat quality index defined as a combination of ultimate pH, reflectance and water holding capacity measurements (Jacquet *et al.*, 1984), but did not consider each component trait of this index. The aim of this study was to estimate genetic parameters for all elementary technological meat quality traits and their relationships with production traits.

Material and methods

Data

Data analysed were collected between 1988 and 1999 in the French central test stations in Large White (LW, 11500 animals) and French Landrace (FL, 4963 animals) breeds on slaughtered sibs of on-farm tested candidates. Animals (castrates and females) were tested between 35 and 100 kg liveweight. Since 1995, feed intake was individually recorded using ACEMA 48 electronic feeders. Pigs were slaughtered during the week following the final weight measurement and a standardised cutting of one half carcass was performed, as described in Anonymous (1990). Performances for four production traits were individually recorded: average daily gain and food conversion ratio between 35 and 100 kg liveweight (ADG, FCR), dressing percentage of the carcass (ratio of carcass weight to liveweight) (DP), and carcass lean content (combination of the weight of three carcass joints (ham, backfat and loin) expressed as a percentage of the half-carcass weight: $CLC = 5.684 + 1.197$ % ham + 1.076 % loin - 1.059 % backfat (Métayer & Daumas, 1998)).

Three meat quality measurements were taken on the ham 24 hours after slaughter:
- ultimate pH of the Adductor femoris muscle (until 1993 – PHAD) or of the Semimembranosus muscle (since 1993 – PHSM) ;
- reflectance of the Gluteus superficialis muscle at 630 nm using a Valin-David Retrolux reflectometer (until 1993 – REFL) or using a Minolta Chromameter CR-300 (since 1993 - paleness L*) (lower values indicate darker meat);
- water holding capacity (WHC): time for a piece of filter paper to become wet when put on the freshly cut surface of the Gluteus superficialis muscle (higher score = better WHC).

A subjective score (SCOR) of meat quality (PSE note) was also given, taking jointly into account colour, humidity and consistency of the freshly cut ham (higher SCOR = better quality). The pH, reflectance and water holding capacity measurements were combined in two synthetic meat quality indexes (MQI1, MQI2), predictors of the technological yield of cooked Paris ham processing (MQI1 = -35 + 8.329 PHAD + 0.127 WHC – 0.0074 REFL until 1993 (Gueblez et al., 1990) ; MQI2 = -41 + 11.01 PHSM + 0.105 WHC – 0.231 L* since 1993).

Models of Analysis

Variance and covariance components were estimated using a REML procedure applied to multivariate individual animal models, using the VCE(4.2) software. In the 12 % of litters for which several piglets were controlled, one single animal per litter was randomly sampled and considered in the analyses. The model for production traits included the group of contemporary animals (year x station x batch) and sex as fixed effects, the random effect of additive genetic effect for each animal, and weight at the beginning of the test period (ADG, FCR) or carcass weight (DP, CLC) as a covariate. Because of low cell size for contemporary group effect (station x slaughter date) for meat quality traits in each breed, records were pre-adjusted considering the whole data set of slaughtered LW, FL and Piétrain animals using a linear model including the fixed effects of slaughter group, sex and breed, and carcass weight as a covariate. (Co)variance components were then estimated from the residuals of this model using an individual animal model. Three generations of ancestors where considered in the pedigree files.

Results and discussion

Heritability estimates for meat quality traits range from 0.05 to 0.25. They tend to be slightly higher in the LW than in the FL breed (except for WHC), particularly for PHAD, REFL and MQI1. Estimated values for ultimate pH are generally in agreement with literature results (Ducos, 1994, Sellier, 1998). The surprising extremely low heritability of PHAD in FL breed was already obtained by Tribout et al., (1996) and Cole et al., (1988). Heritabilities for reflectance measurements are low, but close (L*) to results of De Vries et al.(1994) and Knapp et al., (1997). Heritability estimates for WHC agree with literature average (0.15 - Sellier, 1998), but are lower than results of De Vries et al., (1994) (0.20 for WHC) and of Hovenier et al., (1992) (0.30 for drip loss). Moreover, the higher heritability of water holding capacity in LW than in FL populations quoted by Knapp et al., (1997) and Ducos (1994) is not confirmed here. Finally, heritability estimated for SCOR in both breeds (0.21) is consistent with average literature results (Sellier, 1988).

Table 1. Estimated heritabilities (diagonal), genetic (above diagonal) and phenotypic (below diagonal) correlations among meat quality traits in the Large White breed.

Traits	PHAD	PHSM	REFL	L*	WHC	SCOR	MQI1	MQI2
PHAD	0.24 a	0.90 b	-0.54 b	-0.91 c	0.44 c	0.23 b	0.94 a	0.74 b
PHSM	-	0.21 a	-0.51 c	-0.67 a	0.54 b	0.79 a	0.62 c	0.92 a
REFL	-0.38	-	0.19 a	0.82 b	0.07 c	-0.48 b	-0.60 b	-0.67 c
L*	-	-0.58	-	0.25 a	-0.40 b	-0.96 a	-0.71 b	-0.80 a
WHC	0.19	0.34	-0.19	-0.24	0.12 a	0.52 b	0.53 b	0.62 a
SCOR	0.25	0.48	-0.34	-0.58	0.44	0.21 a	0.33 b	0.92 a
MQI1	0.95	-	-0.60	-	0.49	0.39	0.25 a	0.75 c
MQI2	-	0.92	-	-0.75	0.53	0.61	-	0.24 a

a: std err≤ 0.04 ; b: 0.05 ≤std err ≤ 0.09 ;c: 0.10 ≤ std err ≤ 0.20 ; d: std err ≥ 0.21

Table 2. Estimated heritabilities (diagonal), genetic (above diagonal) and phenotypic (below diagonal) correlations among meat quality traits in the French Landrace breed.

Traits	PHAD	PHSM	REFL	L*	WHC	SCOR	MQI1	MQI2
PHAD	0.05 a	1.00 a	-0.50 d	-0.97 b	-0.06 d	0.58 c	0.57 c	0.99 c
PHSM	-	0.18 a	-0.48 d	-0.68 b	0.43 b	0.59 b	0.93 a	0.97 a
REFL	-0.35	-	0.12 a	0.82 c	-0.74 c	-0.93 b	-0.80 c	-0.65 c
L*	-	-0.59	-	0.22 a	-0.25 b	-0.79 b	-0.93 a	-0.77 a
WHC	0.22	0.35	-0.28	-0.24	0.15 a	0.66 b	0.54 c	0.52 b
SCOR	0.31	0.50	-0.41	-0.60	0.45	0.21 a	0.86 c	0.76 b
MQI1	0.95	-	-0.63	-	0.55	0.45	0.10 a	0.99 b
MQI2	-	0.93	-	-0.74	0.54	0.63	-	0.20 a

a: std err≤ 0.04 ; b: 0.05 ≤std err ≤ 0.09 ;c: 0.10 ≤ std err ≤ 0.20 ; d: std err ≥ 0.21

With the exception of WHC, which is uncorrelated with PHAD in the FL breed and with REFL in the LW breed, all elementary meat quality traits show moderate to strong favourable genetic correlations, in agreement with literature results (Ducos et al., 1994). Correlations between the two pH and between the two reflectance measurements are very strong (not significantly different from 1). Reflectance (REFL in LF, L* in LW) appears as the most closely correlated trait with SCOR. Genetic correlation estimates between WHC, SCOR, PHSM and L* are similar in both breeds. Relationships of PHAD with both WHC and SCOR, and of REFL with both WHC and SCOR seem to differ greatly between breeds, but the trend of the relationship (positive or negative correlation) is generally the same. Genetic correlations between meat quality indexes and their component traits are strong (in decreasing order of correlation: pH, reflectance and WHC). SCOR is likewise strongly correlated with synthetic indexes, except MQI1 in the LW breed (this latter trait being also less correlated with both reflectance measurements in the LW than in the FL breed).

Genetic correlations between growth rate and meat quality traits are null or slightly favourable (WHC, MQI1, MQI2) in the LW breed (except for REFL), and null (PHAD,

REFL, L*) or slightly unfavourable in the FL breed. No correlation involving DP was found significantly different from zero in either breed, except the slightly unfavourable relation with PHSM in the FL breed (-0.12). This lack of relationship is consistent for ADG but not for DP with the review of Ducos (1994), who reported a low unfavourable trend between DP and meat quality traits.

Except for REFL and PHAD presenting favourable relations with FCR (standard errors being large due to data structure), all traits seems to be unfavourably correlated with feed efficiency (except WHC in LW). This is in agreement with results from Garreau et al., (1998) in the LW breed, Tribout et al., (1996), and the review of Sellier (1998) reporting a general antagonism between FCR and technological meat quality traits, in particular meat colour.

CLC presents a clear negative relation with ultimate pH, and a slight favourable trend with WHC. All other meat quality traits are unfavourably correlated with muscle quantity, the antagonism being more important in the LW breed. As both populations are free from the halothane sensitivity allele, the antagonism between muscle quality and muscle quantity observed here may be explained by the positive correlation between CLC and glycolytic potential, which presents close genetic correlations with ultimate pH, reflectance, and to a lesser extent, with water holding capacity (Larzul et al., 1998).

Table 3. Genetic correlations between production and meat quality traits in the LW breed.

Traits	PHAD	PHSM	REFL	L*	WHC	SCOR	MQI1	MQI2
ADG	0.06 b	0.10 b	0.14 c	0.03 b	0.23 b	0.04 b	0.10 a	0.10 b
FCR	0.27 c	0.20 b	0.16 c	-0.28 b	-0.30 b	0.31 b	0.13 c	0.15 b
DP	0.10 b	-0.05 b	0.14 b	-0.05 b	0.03 b	-0.09 b	0.04 b	-0.03 b
CLC	-0.36 b	-0.24 b	0.18 b	0.36 a	0.16 b	-0.41 b	-0.34 b	-0.22 a

a: std err≤ 0.04 ; b: 0.05 ≤std err ≤ 0.09 ;c: 0.10 ≤ std err ≤ 0.20 ; d : std err ≥ 0.21

Table 4. Genetic correlations between production and meat quality traits in the FL breed.

Traits	PHAD	PHSM	REFL	L*	WHC	SCOR	MQI1	MQI2
ADG	0.05 c	-0.14 c	0.07 c	0.05 b	-0.23 b	-0.18 b	-0.10 c	-0.17 b
FCR	-0.48 d	0.48 b	0.22 c	-0.32 c	0.19 b	0.41 c	-0.04 d	0.41 c
DP	0.13 d	-0.12 a	-0.27 c	0.03 b	0.03 a	0.08 b	-0.02 b	-0.05 b
CLC	-0.38 d	-0.12 a	0.11 b	0.15 b	0.09 b	-0.26 b	-0.17 b	-0.13 a

a: std err≤ 0.04 ; b: 0.05 ≤std err ≤ 0.09 ;c: 0.10 ≤ std err ≤ 0.20 ; d: std err ≥ 0.21

References

Anonymous, 1990. Techni-Porc, 13 (5): 44-45

Cole, G., G. Le Hénaff & P. Sellier, 1988. J. Rech. Porc. France, 20: 249-254

De Vries, A.G., P.G. Van der Wal, T. Long, G. Eikelenboom & J.W. M. Merks, 1994. Livest. Prod. Sci., 40: 277-289

Ducos, A., 1994. Techni-Porc, 173: 35-67

Garreau, H., T. Tribout, & J.P. Bidanel, 1998. Techni-Porc, 21(3): 37-43

Guéblez, R., C. Le Maître, B. Jacquet & P. Zert, P.,1990. J. Rech. Porc. France 22: 89-96

Hermesch, S., B.G. Luxford & H.U. Graser, 1998. Proceedings of the 6th world congress on genetic applied to livestock production, Armidale, Australia, 23: 511-514

Hovenier, R., E. Kanis, T. Van Asseldonk & N.G. Westerink, 1992. Livest. Prod. Sci, 32: 309-321

Knapp, P., A. Willam & J. Sölkner, 1997. Livest. Prod. Sci, 52: 69-73

Larzul, C., P. Le Roy, G. Monin, & P. Sellier, 1998. INRA Prod. Anim, 11(3): 183-197

Métayer, A., & G. Daumas, 1998. J. Rech. Porc. France, 30: 7-11

Sellier, P., 1998. Genetics of Meat and Carcass Traits. In: The Genetics of the Pig, M.F. Rothschild and A. Ruvinsky (editors), Cab International, 463-510

Tribout, T., H. Garreau, & J.P. Bidanel, 1996. J. Rech. Porc. France, 28: 31-38

Dissection of genetic background underlying meat quality traits in swine

J. Szyda[1], E. Grindflek[2], Z. Liu[3] & S. Lien[2]

[1] *Department of Animal Genetics, Agricultural University of Wrocław, Kożuchowska 7, 51-631 Wrocław, Poland*
[2] *Department of Animal Science, Agricultural University of Norway, 1432 Ås, Norway*
[3] *United Datasystems for Animal Production (VIT), Heideweg 1, 27-283 Verden, Germany*

Summary

The primary goal of this study was to localise the quantitative trait loci (QTL) responsible for meat quality in swine. A total of 37 traits were scored on 305 backcross individuals from a cross between Duroc and Norwegian Landrace breeds. Genotype information from chromosomes 4, 6, and 7, was available for generations F1 and F2. Statistical procedures applied for the analysis involved (i) the Lander & Botstein (1989) interval mapping, (ii) the composite interval mapping (Zeng, 1994), (iii) regression method (Haley & Knott, 1992) and (iv) a modified Haley and Knott approach including random animal additive genetic effect and relationship between individuals. For the majority of traits analysed there was no clear evidence on segregating QTL, with the exceptions of subacid flavour, diverging smell and intramuscular fat for which QTL could be localised on chromosomes 4, 7, and 6, respectively.

Keywords: QTL detection, Swine meat quality, Mixed inheritance model, Backcross

Introduction

Experimental designs based on crossing inbred lines, which are divergent for the trait of interest, provide the most informative (i.e. powerful) data available for mapping quantitative trait loci (QTL) in livestock. However in practice, instead of inbred lines, it is common to cross divergent outbred populations or breeds. In such a case, one can still take advantage of a special experimental design, but the degree of heterozygosity and linkage disequilibrium are lower. Furthermore, polygenic values of individuals from the parental generation belonging to the same breed (population) are no longer identical.

Several statistical approaches have been proposed for modelling such experimental data (Lander & Botstein, 1989; Haley & Knott, 1992; Zeng, 1994). Some further enhancements comprise incorporation of effect of family and nonzero covariances between observations (Xu, 1998).

Here, for the analysis of backcross data from a cross between Duroc and Norwegian Landrace breeds, a model based on the approach of Haley and Knott was applied, with modifications including the introduction of a random animal additive genetic effect and covariances between related individuals. Results based on this model are compared with those based on interval (Lander & Botstein, 1989; Haley & Knott, 1992) and composite interval mapping (Zeng, 1994) methods.

Material

Phenotypic data on carcass characteristics, taste panel (16 traits describing taste, smell, texture, and colour), water-holding capacity, fatty acid content, amounts of intramuscular fat, protein in muscle, and connective tissue protein were collected on 305 individuals. The F2 generation animals originated from a cross between Duroc sires and Norwegian Landrace dams in a parental generation, and then from backcrossing each of five F1 sires with eight Landrace-Yorkshire dams. For F1 and F2 individuals marker genotypes at chromosome 4 (11 markers covering 126 cM), chromosome 6 (9 markers covering 134 cM), and chromosome 7 (9 markers covering 137 cM) were available.

Methods

Preliminary analysis without marker information

In order to get a better insight into the characteristics of the data, preliminary analysis of phenotypic observations, ignoring the molecular information, was performed. For each trait, frequency distributions, reflecting probability density functions of phenotypes, were visually checked, particularly for extreme deviation from normality, as each of the methods used for QTL mapping assumes a normal distribution of phenotypes. Furthermore, a sequence of linear regression models was fitted to the data in order to identify which of the available covariates: sire, dam, and sex, have a significant effect on the observed phenotypic variation. Differences in goodness of fit between particular models were assessed using the likelihood ratio test with a number of degrees of freedom corresponding to the difference in the number of parameters. A final step comprised the estimation of additive genetic variance of the analysed traits. Variance components were estimated via restricted maximum likelihood, using the SAS package (SAS, 1996).

QTL detection

To map QTL, four statistical models were applied: Lander & Botstein (1989), Zeng (1994), Haley & Knott (1992), and a modified Haley & Knott (1992) model. The first two models were implemented via the QTL Cartographer package (Basten *et al.*, 1994).

Model
The modification of Haley & Knott approach developed in this study involves incorporation of a random animal additive genetic effect into our statistical model:

$$y_i = \mu + \alpha_i + \beta x_{\beta i} + q_1 x_{q1i} + q_2 x_{q2i} + e_i.$$

Where: y_i represents a quantitative trait value of individual i; μ is the overall mean; α_i is a random additive genetic effect of animal i; β represents nongenetic fixed effects (e.g. sex); q_1 and q_2 denote respectively the fixed effects of a heterozygous (say, Qq), and a homozygous (say, qq) QTL genotype; e_i is the random residual; $x_{\beta i}$, x_{q1i}, and x_{q2i} represent appropriate elements of design vectors for β, q_1, and q_2.

(Co)variance structure
An additional feature of the introduced model is the incorporation of information on relationship between individuals into the (co)variance matrix. While methods of Lander and Botstein, Zeng, and Haley and Knott assume no correlation between phenotypic observations y_i, our approach models trait values of full- and halfsibs as correlated. This imposed correlation is based on the similarities between polygenic effects of related individuals, as described by the numeric relationship matrix (Henderson, 1976).

Results

All of the available traits, in which the frequency distribution did not show especially extreme departures from normality, were subjected to QTL mapping using the Lander and Botstein, and Zeng methods. Based on the preliminary analysis, subacid flavour (SF), diverging smell (DS), and amount of intramuscular fat (IMF), showed the most encouraging evidence for a segregating QTL, and were chosen for further analysis.

For each of the three traits chosen, paternal and maternal effects appeared to be significant sources of variation. Additionally, for IMF a significant effect of sex was observed. The estimated variance components resulted in high heritabilities of: 0.554, 0.426, and 0.387, respectively for SF, DS, and IMF. The heritability estimates are subjected to large standard errors due to a small sample size.

Likelihood profiles based on four statistical models for DS, SF, and IMF are shown in Figure 1.

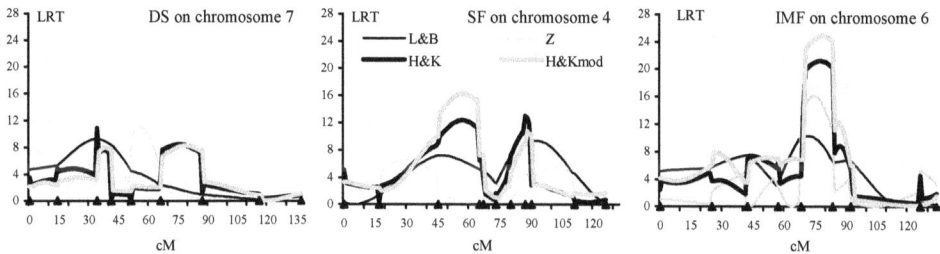

Figure 1. Likelihood profiles from Lander and Botstein (L&B), Zeng (Z), Haley and Knott (H&K), and the modified Haley and Knott (H&Kmod) methods, for diverging smell (DS), subacid flavour (SF), and the amount of intramuscular fat (IMF). Triangles represent positions of markers.

Table 1. Estimates of QTL position (Pos) to the leftmost marker in cM, Likelihood Ratio Test value (LRT), and type I error probability (α), obtained from Lander & Botstein (L&B), Zeng (Z), Haley & Knott (H&K), and the modified Haley & Knott (H&Kmod) methods, for diverging smell (DS), subacid flavour (SF), and the amount of intramuscular fat (IMF).

Model	DS: chromosome 7			SF: chromosome 4						IMF: chromosome 6		
				QTL 1			QTL 2					
	Pos	LRT	α^1	Pos	LRT	α^1	Pos	LRT	α^1	Pos	LRT	α^1
L&B	34	09.38	0.00219	47	07.23	0.00717	870	10.46	0.00122	71	10.28	0.00134
Z	55	11.15	0.00084	29	06.54	0.01055	102	5.72	.01677	74	16.12	0.00006
H&K	34	10.92	0.00425	57	12.38	0.00205	87	13.04	0.00147	76	21.17	0.00003
H&Kmod	79	08.27	0.01600	57	16.27	0.00029	88	10.81	0.00449	78	24.86	≈ 0.0

[1] Type 1 error rates are based on a χ^2 distribution with 1 df for L&B and Z methods, with 2df for H&K and H&Kmod.

For DS, a moderately significant QTL was located on chromosome 7. However, there is an inconsistency in the estimates of QTL position among applied models and, except the approach of Zeng, the evidence of QTL is weaker than in the case of SF and IMF (Table 1).

All four likelihood profiles on chromosome 4 for SF indicate the existence of two linked QTL, with the most probable location within intervals S0001-A-FABP and S0217-S0073, respectively. The highest significance, with type I error rate of practically zero, was observed for IMF on chromosome 6 (Table 1). For IMF, all the applied models point at the interval S0003-S0228 as the most probable QTL location.

Discussion

Three of the meat quality traits recorded in a swine backcross population showed a segregating QTL responsible for a part of their quantitative variation. The most striking evidence was found for the amount of intramuscular fat, with a putative QTL located on chromosome 6. This remains in agreement with recent results of Gerbens *et al.,* (1999) who found an association between IMF and polymorphism at H-FABP marker on the same chromosome.

Results for subacid flavour indicate that two QTL linked on chromosome 4 are segregating. There is however a discrepancy between interval mapping methods and a composite intervals mapping approach related to the estimated position of the left QTL. The latter approach localises both QTL more apart one from another and provides less significant evidence than the three interval mapping methods. The statistical model underlying composite interval mapping method is especially designed to deal with multiple QTL, while the proper separation of effects of linked QTL is a known drawback of interval mapping models, that is the most plausible explanation for the observed discrepancy between the two groups of methods.

The lowest significance and thus the largest differences between four likelihood profiles are observed for diverging smell on chromosome 7. For reliable statistical inferences concerning a putative QTL location at this chromosome more data would be required.

References

Basten, C.J., B.S. Weir & Z.-B. Zeng, 1994. Computing strategies and software. In: Procee-dings of the 5th World Congress on Genetic Applied to Livestock Production, C. Smith, J.S. Gavora, B. Benkel, J. Chesnais, W. Fairfull, J.P. Gibson, B.W. Kennedy & E.B. Burnside (editors), Organizing Committee, 5th World Congress on Genetics Applied to Livestock Production, Guelph, Ontario, Canada, 22: 65-66

Gerbens, F., A.J. van Erp, F.L. Harders, F.J. Verburg, T.H. Meuwissen, J.H. Veerkamp & M.F, 1999. Effect of genetic variants of the heart fatty acid-binding protein gene on intramuscular fat and performance traits in pigs. J. Anim. Sci. 77: 846-52

Haley, C.S. & S.A. Knott, 1992. A simple regression method for mapping quantitative trait loci in line crosses using flanking markers. Heredity 69: 315-324

Henderson, C.R., 1976. A simple method for computing the inverse of a numerator relation-ship matrix used in the prediction of breeding values. Biometrics 32: 69-83

Lander, E.S. & D. Botstein, 1989. Mapping Mendelian factors underlying quantitative traits using RFLP linkage maps. Genetics 121: 185-199

SAS Institute Inc., 1996. SAS/STAT Software: Changes and Enhancements through Release 6.11, SAS Institute Inc., Cary, NC.

Xu, S., 1998. Mapping quantitative trait loci using multiple families of line crosses. Genetics 148: 517-524

Zeng, Z.-B., 1994. Precision mapping of quantitative trait loci. Genetics 136: 1457-1468

Skeletal muscle fibres as factors for pork quality

A. H. Karlsson[1], R. E. Klont [2] & X. Fernandez[3]

[1]*Danish Institute of Agriculture Sciences, P.O. Box 50, DK-8830 Tjele, Denmark*
[2] *ID-DLO, P.O. Box 65 NL-8200 AB Lelystad, The Netherlands*
[3]*INRA de Theix, FR-63122 Saint Genès Champanelle, France*

Summary

Interaction between muscle fibres, perimortem energy metabolism and different environmental factors determine *post mortem* transformation from muscle to meat. Muscle fibres are not static structures, but easily adapt to altered functional demands, hormonal signals, and changes in neural input. Their dynamic nature makes it difficult to categorise them into distinct units. It must be realised that some properties may change without affecting others or without changing histochemical appearance of a given fibre such as species specificity, developmental or adaptive processes or pathological conditions, and that transient fibre types exist. Therefore, a distinction between specific types must strictly refer to the method that has been used for the typing. Theoretically, there will be at least as many fibre types as there are motor units in a muscle, and within a muscle a fibre type may show a continuum of structural and functional properties which overlap with other fibre types.

Histochemical and biochemical properties of a muscle, such as fibre type composition, fibre area, oxidative and glycolytic capacities, and glycogen and lipid contents, are factors that have been found to influence meat quality. An important factor for *post mortem* changes and meat quality is the metabolic response that takes place in the different fibre types *pre-slaughter*. Selection for leaner pigs and for a higher proportion of large muscle fibres, especially of type IIB, can result in poor capillarisation and consequently an insufficient delivery of oxygen and substrates and elimination of end products, such as CO_2 and lactate, and thereby a reduced pork quality.

In the future, fibre types in pig muscle will probably be investigated with more advanced and sensitive techniques, which makes it is possible to look at adaptations in different contractile proteins, sarcoplasmic proteins as well as other muscle proteins that are of importance for cell differentiation and muscle cell metabolism. Furthermore, single fibre dissection and quantitative biochemical analyses may increase the knowledge about metabolic and contractile properties of the muscle fibre.

The literature indicate possibilities to include muscle fibre characteristics in breeding schemes for improved meat quality, while preserving optimal production traits. In order to use muscle fibre characteristics in a beneficial way for future breeding programmes, further investigations are needed to better understand the physiological mechanisms. Selection experiments based on biochemical and histochemical characteristics determined in biopsies or otherwise and study of the correlated selection responses, may possibly provide better tools to study these relationships.

Keywords: Pigs; Pork, Meat; Quality; Muscle fibre

Reprinted from Livestock Production Science Vol. 60, A.H. Karlsson, R.E., Klont, X. Fernandez, Skeletal muscle fibres as factors for pork quality, 255-269, Copyright (1999), with permission from Elsevier Science.

Introduction

Meat quality shows a large variation both within and between animals and even within distinct muscles. There are many factors that contribute to this variation such as species, breed, genotype, sex, age, nutrition and slaughter treatment. Contractile and metabolic properties of skeletal muscle may strongly affect the pattern of energy metabolism in live animals (see review by Hocquette *et al.*, 1998), as well as during the *post mortem* conversion of muscle to meat (i.e. the onset of *rigor mortis*) (Monin & Ouali, 1991). Physiological characteristics of skeletal muscles also explain the variability in metabolic responses of muscle tissue during pre-slaughter stress, and subsequent rate and extent of *post mortem* pH fall. Therefore, a strong link exists between muscle energy metabolism in live animals, metabolic response to slaughter stress and meat qualities.

Research on the relationships between skeletal muscle characteristics and meat quality is important for improving our understanding of the molecular basis of the phenotypic expression in skeletal muscle and its interaction with different environmental factors. The aim of the present report is to review present knowledge on the relationship between adult skeletal muscle fibre type compositions and pork quality.

Skeletal muscle function, structure and composition

A unique characteristic of skeletal muscle is its diversity, which is created by its design, its fibre type, or cell type, composition and the heterogeneity of individual fibres (cells). We know that no skeletal muscle within an animal is identical with a second. Homologous muscles exhibit differences in fibre composition between species and between strains. Striated skeletal muscles are responsible for body movement, and they are composed of multinucleated muscle fibre cells, may extend the entire length of the muscle and are arranged in myofibrils. Each myofibril is constructed of two types of longitudinal filaments. The thick type, contains mainly the protein myosin. Myosin consists of two identical heavy chains and two pairs of light chains. Small globular projections at one of the heavy chains form the heads, which have ATP binding sites as well as an enzymatic capacity to hydrolyse ATP. The thin filaments contain the proteins actin, tropomyosin, and troponin.

The main components of muscle tissue are water, protein and lipid. Lean muscle tissue from a typical porcine muscle after *rigor mortis* consists on average of 75 percent water, 22 percent protein and varying amounts of lipid and carbohydrates (Monin *et al.*, 1986). The muscle proteins can be classified into three main groups: myofibrillar, sarcoplasmic, and stroma proteins. The myofibrillar proteins constitute the largest fraction and represent about 60 percent of the total protein content. The sarcoplasmic proteins represent about 30 percent of the total proteins. This fraction consists of myoglobin, which is a muscle pigment, and many different enzymes involved in energy metabolism. The stroma proteins represent about 10 percent of all proteins. The main components of the stroma proteins are collagen and elastin, which are the main components of the connective tissue.

Muscle fibre types – Classification systems

Histochemical staining with Sudan black-B has been used to classify fibres into red and white fibres (Beecher *et al.*, 1965; Beecher *et al.*, 1969). On the basis of histochemical staining of myosin ATPase pre-incubated at pH 9.4 observations of two distinct types of muscle fibre (types I and II) has been done in both humans (Engel, 1962) and in guinea pigs (Robbins *et al.*, 1969). Type I fibres had a high content of mitochondrial enzymes and were low in myofibrillar ATPase and phosphorylase activity. The converse pattern was seen for type II fibres. Brooke & Kaiser (1970) grouped the type II fibres into three subtypes according to ATPase staining after acidic pre-incubation: IIA, IIB, IIC (Tab.1). A linear relationship has been found between the activity of myosin ATPase and the contraction rate of muscle fibre (Burke et al., 1973; Garnett et al., 1978).

Table 1. Properties of muscle fibres classified according to the system of Brook and Kaiser.

Fibre type	I	IIA	IIB
Contractile properties			
Myosin ATPase	Slow	Fast	Fast
Metabolic properties			
Glycolytic capacity	++	++(+)	++(+)
Oxidative capacity	+++	++(+)	+
Glycogen content	++	++(+)	++(+)
Triglyceride content	+++	++	(+)
Capillary density	+++	++(+)	++

The oxidative capacity of muscle fibres can be assessed by staining for the enzymes NADH-tetrazolium reductase (NADH-TR; Novikoff et al., 1961) or succinate dehydrogenase (SDH; Ogata, 1958). A combination of staining for myosin ATPase at pH 9.4 and NADH-TR, or SDH, separates the fibres into slow-twitch high-oxidative (SO), fast-twitch high-oxidative (FOG), and fast-twitch low-oxidative (FG). These fibre types are also referred to as ß-red, α-red, and α-white, respectively. The system by Brooke & Kaiser (1970), concerning type II fibres, corresponds only to a minor degree to the combined system of ATPase and NADH-TR or SDH, where there is an overlap between fibres of type IIA and IIB (Pierobon-Bormioli et al., 1981; Hintz et al., 1985; Lind & Kernell, 1991). There is a close agreement between type I and β-red fibres. Kiessling & Hansson (1983) found close agreement between the fibre types β-red, α-red and α-white, and type I, IIA, and IIB, respectively, in nine porcine muscles. Essén-Gustavsson & Lindholm (1984) stained various pig muscles for ATPase (types I, IIA, IIB) and NADH-TR, and classified each fibre type as dark, medium or low, based on a combination of the two staining techniques ATPase and NADH-TR. These authors found that 15-20 percent of type IIB fibres in M. longissimus dorsi stained medium for NADH-TR, and these would therefore be classified as α-red (Figure1).

Comparison of different muscle fibre types using different classification systems
(Essén-Gustavsson, 1996)

•Types: **I, SO, ST** and **↑R** are identical and interchangeable.

•Types: **IIA-IIB** or **FOG-FG, FTH-FT, ↠w-↑R** has a certain overlap.

Figure 1. Comparisons of different muscle fibre types.

There is a correlation between fibre types classified on the basis of pH lability of myofibrillar ATPase reactivity according to Brooke & Kaiser (1970) on one hand, and on the other hand, the myosin heavy chain (MHC) isoforms expressed by muscle fibres (Termin *et al.*, 1989; Gorza, 1990; Fig. 2). Histochemically classified type I fibres express MHC-1 isoform, type IIA express MHC-IIA isoform, and type IIB fibres express either MHC-IIB or the MHC-IIX isoforms (Bar & Pette, 1988; Schiaffino *et al.*, 1989; Gorza, 1990). It has been shown that maximum contraction rate of a fibre correlate with the MHC isoforms expressed (Schiaffino *et al.*, 1988).

The fibre type called type IIC (Kugelberg, 1976; Lutz *et al.*, 1979; Billeter *et al.*, 1980), or type IIX (Schiaffino *et al.*, 1989), is said to be a transition form between types I and II and it can be seen in neonatal individuals and during repair from a muscle injury (Pette & Staron 1990). In principle each muscle fibre is unique, independent of the type it has been categorised into, and may show a continuum of structural and functional properties which overlap with other fibre types.

Fibre functional properties vs. Myosin Heavy Chain isoforms

Functional properties:

- Peak force
- Contraction velocity
- Fatigue resistance
- Oxidative capacity
- Glycolytic capacity
- Myosin ATPase activity

Muscle functional properties are closely related to the Myosin Heavy Chain (MHC) isoform.

Immunohistochemical methods using MHC would therefore be a more sensitive approach.

Figure 2. Fibre functional properties vs. MHC isoforms.

Factors influencing muscle fibre type composition and fibre morphology

Sex

There is not much literature published concerning the effect of sex on muscle fibre composition. Miller *et al.* (1975) did not find any sex differences in fibre composition, and Staun (1963) did not find any sex differences in fibre number or fibre diameter. Karlsson (1993) found that the cross-sectional area of fibre types I, IIA and IIB were smaller in *M. longissimus dorsi* in entire male pigs compared to gilts. Petersen *et al.* (1998) found smaller type IIA and type IIB fibres in boars compared to female pigs, and because the weight of the *M. longissimus dorsi* was similar, this indicates that the number of muscle fibres is higher in male pigs.

Age

Muscle fibres are oxidative, at birth (Moody *et al.*, 1978). The age at which three fibre types can be distinguished histochemically from each other, based on the myosin ATPase method, ranges from one (Moody *et al.*, 1978) to four weeks (Cooper *et al.*, 1970; Lefaucheur & Vigneron, 1986). The proportion of oxidative fibres decreases, while the proportion of glycolytic fibres increases, in *M. longissimus dorsi* during growth (Cooper *et al.*, 1970; van den Hende *et al.*, 1972; Szentkuti & Cassens, 1978; Schlegel, 1982; Kiessling *et al.*, 1982; Bader, 1983; Salomon *et al.*, 1983, Lefaucheur & Vigneron, 1986, Wegner & Ender, 1990, Solomon *et al.*, 1990). According to Swatland (1978) the proportion of glycolytic fibres increases in *rectur femoris* during growth. Suzuki & Cassens (1980) studied pigs from birth to eight weeks of age and they found that the proportion of type I fibres increased in several porcine muscles.

The cross-sectional area of all fibres increases during growth (Chrystall *et al.*, 1969; Hegarty *et al.*, 1973). Chrystall *et al.* (1969) found in porcine *M. longissimus dorsi* that muscle fibre diameter increased 100 percent from birth to 25 days, whereas from 100 to 125 days of age, muscle fibre diameter increased only 10 percent. The rate of muscle fibre growth after 150 days of age was very slow. In modern pigs, it was shown that the rate of increase in muscle fibre area was almost constant from 25 to 90 kg liveweight, and that the rate of growth was two-fold in IIB fibres compared to type I (Oksbjerg *et al.*, 1994a). The size of various fibre types in *M. longissimus dorsi* increased at different rates. The diameter of type II fibres increased faster than that of type I (Cooper *et al.*, 1970; Davies, 1972; Rede *et al.*, 1986; Wegner & Ender, 1990; Fiedler *et al.*, 1991; Oksbjerg *et al.*, 1994a). Research on muscle metabolism of the growing pig has indicated that oxidative capacity is maintained or declines, while glycolytic capacity is enhanced with age (Cooper *et al.*, 1971; Dalrymple *et al.*, 1973; Lefaucheur & Vigneron, 1986; Oksbjerg *et al*, 1994b). According to Oksbjerg *et al.* (1994[a]), the capillary density decreased during growth.

Muscle type

Muscles differ on the basis of fibre composition, rate of fibre differentiation and rate of muscle growth. Most muscles have a fibre composition which is a mix between light and dark fibres. Dark muscles contain predominantly type I and type IIA muscle fibres, while light muscles contain primarily type IIB fibres. Examples of porcine muscles containing a high degree of IIB fibres are *M. longissimus dorsi* (Kiessling and Hansson, 1983; Essen-Gustavsson and Fjelkner-Modig, 1985; Monin *et al.*, 1985), *M. gluteus medius* (Essén-

Gustavsson and Fjelkner-Modig, 1985; *M. rectrus femoris* (Bader, 1982), *M. biceps femoris* (Barton-Gade, 1981), *M. quadriceps femoris* (Barton-Gade, 1981), *M. vastus lateralis* (Kiessling & Hansson, 1983), and *M. semimembranosus* (Barton-Gade, 1981; Monin *et al.*, 1987). Examples of muscles containing a high degree of I and IIA fibres are *M. masseter* and *M. trapezius* (Monin *et al.*, 1987), *M. vastus intermedius* (Bader, 1982), *M. triceps brachii* (Kiessling & Hansson, 1983), *M. infra spinaturs* and *M. supra spinam* (Ruusunen, 1989).

Hormones

It has been found that fibre types can be influenced and regulated by different hormones. Oksbjerg *et al.* (1990, 1994[a]) and Oksbjerg & Sørensen (1996)) found that β-agonists increased fibre area and also caused a IIA to IIB conversion. By treating pigs with porcine growth hormone (pGH) Sørensen *et al.* (1996) found an increased fibre cross-sectional area in the *M. longissimus dorsi*, without affecting the fibre type frequency by the fibre types. However, breed types with relatively large muscle fibres are less responsive to pGH treatment. Insulin deficiency has been found to cause selective atrophy in type II fibres (Armstrong *et al.*, 1975). Nwoye *et al.* (1982) investigated the effect of thyroid hormone on muscle properties. They found that hypothyroidism caused a type II to type I changes in fibre type composition, and vice versa changes were found in hyperthyroidism.

Breed

Muscle fibre differences between pig breeds have been found. In two early studies, muscle fibre diameter differed between breeds (Rubli, 1931; Mauch & Marinesco, 1934). Mean fibre areas of both NADH-TR positive (oxidative metabolism) and NADH-TR negative (glycolytic metabolism) fibres in *M. longissimus dorsi* have been found to be larger in Poland China pigs than in Chester White pigs (Sair *et al.*, 1972). Rede *et al.* (1986) compared highly selected breeds (Large White and Swedish Landrace) with the primitive pig breeds Manguâlitza and Black Slavonic. They found in *M. longissimus dorsi* of selected breeds larger proportions of what they termed 'white muscle fibres' which had a larger diameter. Essén-Gustavsson & Fjelkner-Modig (1985) found that *M. longissimus dorsi* and *M. gluteus medius* of Hampshire pigs had a greater oxidative capacity, a lower glycolytic capacity, and a higher concentration of both glycogen and triglycerides, than those of the Swedish Yorkshire. The corresponding values for the Swedish Landrace pigs were intermediate Hampshire and Swedish Yorkshire.

Halothane genotype

Differences in muscle fibre proportions have been seen between pigs of different halothane genotypes. Essén-Gustavsson & Lindholm (1984) found that halothane-sensitive Swedish Landrace pigs had a lower oxidative capacity and a higher glycogen content in *M. longissimus dorsi* than pigs not reacting to halothane. Lundström *et al.* (1989) investigated the metabolism in *M. longissimus dorsi* and *M. quadriceps femoris* (*rectus femoris*) from the three halothane-genotypes (Hal^{NN}, Hal^{Nn} and Hal^{nn}), but no differences in oxidative or glycolytic enzyme activities were found. Biochemically measured muscle metabolites differed in samples from Hal^{NN} vs. Hal^{nn} pigs, with higher lactate and glucose-6-phosphate and lower glycogen, creatine phosphate and ATP in the latter. The heterozygote pigs (Hal^{Nn}) were between or close to either of the homozygotes. Klont *et al.* (1993) found higher muscle lactate concentrations in Hal^{nn} and Hal^{Nn} pigs compared with Hal^{NN} pigs during anaesthesia. Essén-Gustavsson *et al.* (1992) investigated the muscle fibre characteristics at slaughter in *M.*

longissimus dorsi from pigs of the two homozygote halothane genotypes. No difference in fibre type composition was found, but the Hal[nn]-genotype had larger mean fibre cross-sectional areas

Wild boar vs. Domestic pigs

Differences in muscle fibre characteristics have also been found between the wild boar and the domestic pig. In a study performed by Rede *et al.* (1986), both the wild boar and crosses between the wild boar and the domestic pig had higher proportion of type I fibres in *M. longissimus dorsi* than had the domestic pig. Similar results were found by Essén-Gustavsson & Lindholm (1984), muscles from the wild boar had a higher oxidative capacity than those of Swedish Landrace pigs. McCampbell *et al.* (1974) found a higher myoglobin content in the wild boar than in domestic pigs, which would indicate a greater oxidative capacity. Concerning fibre cross-sectional areas, wild boars had the same or larger areas in type I and type II fibres as domestic pigs. (Rahelic & Puac, 1980-81; Szentkuti *et al.* 1981, Bader 1983; Karlström *et al.*, 1995). According to Szentkuti & Schlegel (1985) the muscle fibre composition seems to be genetically determined. In spite of intensive physical exercise on treadmill, Szentkuti & Schlegel (1985) found that domestic pigs had a smaller proportion of type IIA fibres, and a larger proportion of glycolytic type IIB fibres, compared with wild boars that were kept in a pen to restrict physical activity.

Fibre composition in relation to muscle biochemistry and energy metabolism ante and post mortem

Muscle type and energy metabolism in live animals

Muscle energy metabolism relies on fuels from both extra- and intra-muscular sources. Mainly glucose, lactate, non-esterified fatty acids and triglycerides for the former, and triglycerides and glycogen for the latter. The amount of ATP generated from each of these energy sources and the selection of the appropriate substrates for energy production depend, among other factors, on the metabolic type and contractile activity of muscle tissue (Hocquette *et al.*, 1998).

Plasma glucose is the main energy source for muscle tissue. It has been reported that the degree of vascularisation, and therefore glucose availability, is higher in red than in white muscles (Cassens & Cooper, 1971). In resting muscle, in the absence of insulin or with physiological insulin concentrations present, rates of glucose uptake have been shown to be highest in muscles composed primarily of slow-twitch oxidative fibres, followed by fast-twitch oxido-glycolytic and fast-twitch glycolytic fibres (James *et al.*, 1985; Richter *et al.*, 1984). Fibre type differences in the rate of glucose uptake in resting, insulin-stimulated, and contraction stimulated skeletal muscle may be due to differences in the number of fibres, but not the intrinsic activity of glucose transporting proteins (Goodyear *et al.*, 1991).

The ability of muscle tissue to use plasma glucose is an important feature of the metabolic response to stress before slaughter; as the intracellular energy reserve glycogen degradation will depend on the ability of using plasma glucose as a fuel for ATP production in muscle cell. It is generally found, that predominantly fast-twitch muscles contain more glycogen than predominantly slow-twitch ones in laboratory animals (Villa-Moruzzi *et al.*, 1981), cattle (Lacourt & Tarrant, 1985) and pigs (see review by Fernandez & Tornberg, 1991). Yet, the differences in glycogen content between the predominantly fast-twitch glycolytic and the predominantly fast-twitch oxido-glycolytic muscles depends on the species considered. In sheep for instance, the former contain less glycogen than the latter (Briand *et al.*, 1981) while the reverse is observed in pigs (Monin *et al.*, 1987). For a given muscle, difference in glycogen level can be found between fibre types. Higher glycogen levels have been reported in fast-twitch compared to slow-twitch fibres in cattle (Lacourt & Tarrant, 1985) and pigs (Fernandez *et al.*, 1994b). In pigs, Fernandez *et al.* (1994b) found only slight differences between fast-twitch glycolytic and fast-twitch oxido-glycolytic fibres.

Phosphorous compounds such as ATP and phosphocreatine are very labile and it is difficult to compare *in vivo* levels of these compounds between muscles. Nevertheless, in a review concerning laboratory animals, Kelso *et al.* (1987) noted that, overall, ATP and phosphocreatine levels are higher in predominantly fast-twitch muscles than in predominantly slow-twitch ones. In rabbit muscle using 31P-NMR, Renou *et al.* (1986) reported higher levels of ATP and phosphocreatine, and lower levels of inorganic phosphate in fast-twitch than in slow-twitch muscle.

Pre-slaughter fibre type activation and composition

Pre-slaughter handling of meat animals may lead to physical exercise of different intensity. The nature of the physical activity determines the muscles which are activated according to their anatomical location and function. Within an active muscle, the level of activity will determine which fibre type is activated.

A motor neurone, which is the nerve cell whose axons innervate skeletal muscle fibres, plus the muscle fibres it innervates is called a motor unit. All fibres within a motor unit are of similar metabolic and contractile properties. The motor nerves for the different fibre types have various threshold for activation. The difference is due, at least in part, to the size of the motor neurone, the smallest neurones having the lowest threshold. The small motor neurone innervate type I fibres, and successively larger motor neurones innervate types IIA and IIB fibres. At low level of excitation, some type I fibres will be activated. As the level of excitation increase, *i.e.* the intensity of physical activity, a larger proportion of type I fibres are activated and successively IIA and IIB fibres contract.

Muscle type and adrenaline-induced glycogenolysis

Catecholamines, and more specifically adrenaline, constitute the main category of hormones able to trigger short-term mobilisation of muscle energy reserves during stress responses, through their direct glycogenolytic action on skeletal muscle (Kuchel, 1991). Adrenaline administration induces glycogen depletion in conscious pigs (e.g. Hedrick *et al.,* 1964; Hatton *et al.,* 1972). Nevertheless, the potent action of adrenaline on glycogen metabolism in resting muscle remains unclear. Studies in humans and rats have shown that the glycogenolytic effect of adrenaline is very low in non contracting muscles (Chasiotis *et al.,* 1983). Despite the high rate of conversion of phosphorylase b to a, glycogenolytic rate remains low in resting muscle due to the low level of inorganic phosphate (Pi) at the active site of the enzyme (Pi concentration is far below the corresponding km of phosphorylase a). When the muscle contracts, however, the increased Pi level due to ATP hydrolysis activates phosphorylase a and thereby glycogenolysis.

Under resting conditions, the influence of muscle type on glycogen depletion in response to adrenaline is controversial. Some authors report a higher sensitivity of predominantly slow-twitch muscles (Gorski, 1978; Kennedy *et al.*, 1986, in rats; Lacourt & Tarrant, 1985 in cattle), attributed to a higher degree of vascularisation. By contrast, some authors report a higher sensitivity of predominantly fast-twitch glycolytic muscles (Chasiotis,

1985) and in a given muscle, of the fast-twitch glycolytic fibres (Jensen *et al.*, 1989). In anaesthetized pigs, Fernandez *et al.* (1995a) reported a higher glycogenolytic effect of adrenaline perfusion in the predominantly oxidative *M. tapezius* than in the predominantly glycolytic *M. longissimus dorsi*.

In contracting muscles, and during an exercise of moderate intensity, adrenaline administration increases the glycogenolytic rate in predominantly fast-twitch glycolytic muscles but not in predominantly slow-twitch oxidative muscles (Gorski, 1978; Richter *et al.*, 1982). During an exercise of high intensity, the reverse is observed since the glycogenolytic rate in fast-twitch glycolytic fibres is already at its highest level and cannot be further enhanced by adrenaline (Richter *et al.*, 1982; Greenhaff *et al.*, 1991).

Under practical conditions, pigs to be slaughtered are deprived of food about 12 to 18 h before leaving the farm in order to decrease mortality rate during transport and to reduce *post mortem* pH decline which resulted in a high ultimate pH and thereby improving the technological qualities of the meat. In pigs, Fernandez *et al.* (1994) showed that a 24 h food deprivation induced a significant decrease in glycogen level in the predominantly oxidative *M. semispinalis* (SS) but not in the predominantly glycolytic *M. longissimus dorsi* (LD). In SS muscle, glycogen depletion occurred firstly in slow-twitch oxidative fibres. When food deprivation was prolonged from 24 to 72 h, glycogen was depleted in fast-twitch fibres. In the LD muscle, only fast-twitch fibres showed glycogen depletion after 24 h of food deprivation, but this effect was of lower magnitude than in the SS. The higher sensitivity of predominantly oxidative muscles to food deprivation, under resting conditions, may be attributed to i) their higher chronic activity due to anatomical functions, than in glycolytic muscles and ii) the decrease ability to use plasma glucose during food deprivation due to insulin deficiency.

Mixing unfamiliar pigs on farm, during transport and/or lairage at the abattoir leads to aggressive interactions, *i.e.* increased muscular activity, accompanied by increased circulating catecholamines (*e.g.* Fernandez *et al.*, 1994a). Such behavioural stress triggers glycogen depletion in muscles, but this effect depends on both the anatomical location and the type of muscle. In pigs, aggressive interactions were shown to induce glycogen depletion in the SS muscle, implicated in head movements, but not in the LD muscle (Fernandez *et al.*, 1994b). Furthermore, within the SS muscles, glycogen depletion occurred only in fast-twitch fibres. There was a strong negative correlation between the level of glycogen in fast-twitch fibres and the number of aggressive acts performed by the animals whereas no relationships were found between glycogen and catecholamine levels (Fernandez *et al.*, 1994b). It was further demonstrated that food deprivation enhanced glycogen depletion in the LD in response to aggressive interactions, but this effect was significant only in fast-twitch glycolytic fibres (Fernandez *et al.*, 1995b). The authors concluded that fasting-induced changes in glucose metabolism lead to higher dependency on intracellular energy reserves in working muscles, and more specifically in the most active fibres during this type of stress, namely the fast-twitch fibres.

Fast-twitch muscles generally contain more glycogen than predominantly slow-twitch ones. Subsequently, ultimate pH is usually lower in the former (see review by Fernandez & Tornberg, 1991). Laborde *et al.* (1985) demonstrated a negative relationship between ATPase activity, a marker of contraction speed, and ultimate pH among various pig muscles. *Post mortem* rate of pH decrease depends on the speed of ATP hydrolysis and thus, on ATPase activity (Bendall, 1973). Nevertheless, the relationships between muscle ATPase activity in live animals and *post mortem* rate of pH decline are not clear. In rabbit, Renou *et al.* (1986) showed that predominantly fast-twitch muscles had a higher ATPase activity and thus a higher *post mortem* turnover of ATP, but the rate of pH decline in the corresponding muscles did not differ markedly. This could be attributed to the higher buffering capacity of fast-twitch muscles compared to slow-twitch ones (Rao & Gault, 1979).

Heritability and selection for fibre type characteristics

Traditional pig breeding programs have been based on selection for high growth rate, low feed conversion and for high lean meat percentage. At present, there is a growing interest for including meat quality traits into the breeding goals (de Vries et al. 1998). Several studies have estimated heritabilities and correlation coefficients between meat quality traits (Cameron, 1990; de Vries et al., 1994; Sellier, 1988). Muscle fibre traits have moderate to high heritabilities in pigs (Dietl et al., 1993; Larzul et al., 1997). Henckel et al. (1997) found a positive correlation between muscle gain and both the activity of the oxidative enzyme citrate synthetase and the number of capillaries per fibre. Selection possibilities have been suggested on the basis of different muscle properties in order to improve i.e. the oxidative capacity of muscles and thereby pork quality (Von Lengerken et al., 1994; Maltin et al., 1997). Maltin et al. (1997) compared LD muscle fibre characteristics of pigs from eight different breeding populations in relation to the variation in eating quality. They found that the fibre diameter of fast twitch oxidative glycolytic fibres significantly contributed to the variation in instrumentally determined meat tenderness. Karlsson et al. (1993) found a negative correlation between the proportion of fast twitch (type IIB) fibres and shear force value in the LD muscles from halothane gene free Yorkshire pigs fed on a low protein diet.

Selection of heavily muscled pigs with a low backfat thickness may have encouraged the development of muscle fibres with a large diameter specialised in anaerobic metabolism, which results in a higher capacity for the rapid production of lactate after slaughter (Ashmore & Doer, 1971; Swatland, 1976). This resulted in more type IIB muscle fibres with a larger cross-sectional area in domestic pig breeds compared to native breeds (Weiler et al. 1995).

In studies using modern pig breeds, different trends in muscle fibre characteristics can be observed. Petersen et al. (1997) compared muscle fibre properties of two groups of Danish Landrace pigs, of which one group have been subjected to 20 years selection for growth performance and carcass traits and the other had been kept genetically stable. The study concluded that selection had indeed caused a lower proportion of oxidative type I fibres, but did not result in a larger mean fibre area as determined in longissimus muscles. Larzul et al. (1997) studied phenotypic and genetic correlations of longissimus muscle fibre characteristics in French Large White pigs and concluded that these traits have medium heritability and significant genetic correlations with meat and carcass quality characteristics. Based on their results it be can concluded that an increase in lean percentage would coincide with a decrease in glycolytic fibre proportion and an increase in mean fibre diameter. Brocks et al. (1999) used two lines of Dutch Large White pigs either selected for low backfat thickness or increased growth rate. Lean percentage differed between these two lines (Sonesson et al., 1998). Based on these studies increased leanness appeared to be more associated with muscles containing a low percentage of oxidative (type I and IIA) and a high percentage of glycolytic (type IIB) fibres. This was, however, shown to be muscle and generation dependent, while no significant changes in fibre diameter were observed. Cameron et al. (1997) studied muscle fibre characteristics in pigs that were selected for efficient lean growth rate for seven generations. The selection resulted in increased carcass lean by increasing lean tissue growth rate, but there was no increase in the frequency or area of type IIB fibres.

Relationships between meat quality traits and muscle fibre properties

Meat quality can be defined as 'the total degree of satisfaction that a meat gives the consumer' (Jul & Zeuthen, 1981). Meat quality can be assessed by measuring biophysical and chemical properties, such as water-holding capacity, colour and light reflectance, pH, pigment content, shear force, intra-muscular fat content, and protein extractability. Quality characteristics of particular interest in pork are levels and variability in colour and water holding (Somers & Tarrant, 1984).

After slaughter, glycogen is converted into lactic acid which is accumulated in the muscle. A rapidly increased concentration of lactic acid in a still-warm muscle results in more or less extensive protein denaturation and alters the biophysical properties of meat (the water-holding of the proteins; Bendall & Wismer-Pedersen, 1962). Extensive protein denaturation can lead to PSE meat. (Wismer-Pedersen, 1959; Bendall & Wismer-Pedersen, 1962; Barton-Gade, 1981; Lopez-Bote et al., 1989).

If there is a high frequency of glycogen depleted fibres at slaughter, especially of type IIB, this will influence meat quality. In halothane-gene-free pigs (Hal[NN]) the proportion of glycogen depleted type IIB fibres in M. longissimus dorsi has been found to be positively correlated to ultimate pH and to pH at exsanguination, and negatively correlated to drip loss and light colour (Karlsson et al., 1992). When 30 percent or more of the type IIB fibres were depleted of glycogen, muscles showed a tendency to become dry, firm and dark (DFD). Pigs that are homozygote for the recessive halothane gene have a poor meat quality and a higher frequency of PSE-meat compared to pigs free of this gene, while muscle fibre type composition and muscle enzyme activities were similar (Lundström et al., 1989). The Hal[nn] pigs showed a larger proportion of glycogen depleted type IIB fibres at slaughter, even though all pigs were raised, transported and handled in a similar manner. In the study of Karlsson et al. (1992), the proportion of glycogen depleted type IIB fibres was negatively correlated to pH at exsanguination and positively to drip loss and meat colour. The metabolic response in the Hal[nn]-genotype is related not only to stress-situations prior to slaughter, but also to the fact that these pigs have larger muscle fibre areas and a lower capillarisation of the muscle (Essén-Gustavsson et al., 1992; Fig. 3).

Ultimate meat quality has been predicted from glycogen depletion patterns analysed on samples from M. longissimus dorsi taken two hours post slaughter (Severini et al., 1989). Muscle samples having a high initial pH and numerous fibres staining for glycogen content (PAS-positive fibres), were predicted to give meat of good ultimate quality. Muscle samples which had a large proportion of PAS-negative fibres were expected to give meat of DFD quality with a high initial pH, or meat of PSE quality with a low initial pH-value.

```
┌─────────────────────────────────────────────┐
│           Pork Quality vs.                    │
│         Muscle Fibre Types                    │
│  ───────────────────────────────────────      │
│  •No effect of fibre type composition on the  │
│   lightness (L*)                              │
│  •High percentage of type I fibres in number  │
│   of area increases the redness (a*).         │
│  •High capillary density reduces pH_u, L* and │
│   a*.                                         │
│  •Large CSA of IIB fibres increase WHC.       │
│  ───────────────────────────────────────      │
└─────────────────────────────────────────────┘
```

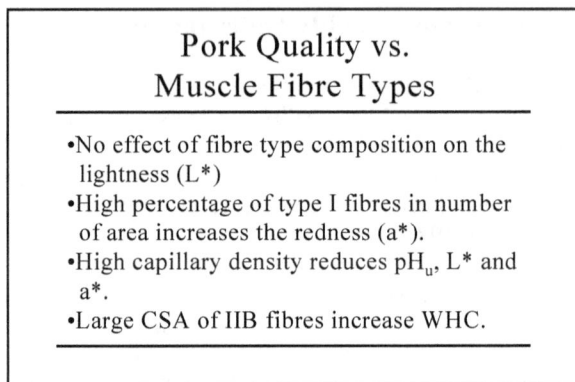

Figure 3. Relationships between fibre types and pork quality.

The classification of the muscle samples at 24 hours *post mortem*, was in good agreement with the prediction made two hours after slaughter, based on initial pH-value and fibre glycogen content, based on PAS-staining intensities.

Fat in muscle is also a factor that may be of importance for meat quality. In sensory tests, meat from fat type pigs obtained high tenderness scores, which may have been due to the large fat content (Rede *et al.*, 1986). Meat from half-Chinese crossbred pigs has been judged to be more tender, juicy and tasty than meat from purebred European pigs (Touraille *et al.*, 1989). Chinese purebred pigs have received higher sensory scores than have Landrace and Duroc (Suzuki *et al.*, 1991). Pigs in both these studies with high sensory scores had more intra-muscular fat (IMF) than the other pigs. Intra-muscular lipid content varies little with metabolic type of muscle (Leseigneur-Meynier & Gardemer, 1991). However, IMF represents the lipid that is stored both intracellular and intercellular and the lipids stored between muscle fascicles. Domestic pigs fed a low-protein diet, had a higher IMF content and lower shear force values in M. *longissimus dorsi* than pigs fed a high-protein diet (Essén-Gustavsson *et al.*, 1992). Muscle fibre composition and the oxidative and glycolytic capacity did not differ between the pigs given the two different diets. IMF values correlated to triglyceride content in the muscle, which in turn was negatively related to mean fibre area. The histochemical staining showed that all type I fibres contained neutral lipids, whereas only about 26 percent of type IIA fibres, and 1 percent of type IIB fibres contained neutral lipids. High triglyceride content in muscle fibres may be one factor of importance for meat quality. Tenderness of meat was lower in Yorkshire pigs compared to Swedish Landrace pigs and Hampshire pigs. The latter pigs had muscles with higher oxidative and lower glycolytic capacity and a higher triglyceride content than muscles from Yorkshire pigs (Essén-Gustavsson & Fjelkner-Modig, 1985). Positive ratings for sensory properties and especially tenderness may thus be related to the oxidative capacity and fat content of muscle fibres. Larzul *et al.* (1997) did not find any relationship between IMF content and the numerical percentage of fibre types at a commercial slaughter weight of purebred Large White pigs and suggested that both traits can be manipulated separately.

Discussion

Early histochemical studies have indicated that there may exist more than three major fibre types. However, the resolution of commonly used histochemical methods seems not to be sufficient to precisely define biochemical fibre type diversities. A better approach to elucidate fibre heterogeneity is offered by immunohistochemical techniques, although they depend on the availability of specific antibodies raised against specific antigens. A more promising technique is single fibre dissection for quantitative microbiochemical or microimmuno-chemical analyses. These microtechniques provide valuable insights in the metabolic enzyme and myofibrilar protein profiles of an individual fibre. The single fibre dissection technique has made it possible to determine enzyme activity profiles biochemically in histochemically typed fibres (Essén et al., 1975), and several studies have shown that great differences exist in the enzyme activity profiles in individual fibres within fibre type populations, which demonstrate a metabolic continuum (Lowry et al., 1980).

Another important information to bear in mind is the number of different fibre types in a muscle. In most studies where the fibre type composition have been analysed, three or four fibre types have been analysed using histochemical staining techniques, such as Brooke & Kaiser (1970) and Peter et al., (1972) on a small sample representing a whole muscle. From results of analyses of enzyme activities on single fibres it appears that singly innervated mammalian fibres within a given motor unit are biochemically similar or identical (Burke et al., 1973), and the variability between fibres within a motor unit is much smaller than between fibres of different units (Nemeth & Wilkinson 1990). Local factors, such as interfascicular and intramuscular location of fibres composing a given motor unit, capillarisation, availability of oxygen and metabolites, active and passive stretch, temperature gradients, may modulate the primary pattern dictated by the neural input, which verifies the dynamic nature of muscle fibres. Therefore, nonuniformity may also exist between fibres within the same motor unit as well as within a single fibre (Staron & Pette, 1987).

A variety of muscle fibre types can be distinguished in a given muscle by different methods, but the number of fibre types is less than the total number of fibres in a given muscle, and if one assumes uniformity of the motor unit, the theoretical number of fibre types within a muscle is equal to the number of motor units (Pette & Vrbodá, 1985). This implies that the phenotypic expression of muscle fibres is primarily dictated by the neuromuscular activity pattern (Pette & Vrbodá, 1985). According to Gauthier et al. (1983) all fibres composing a specific motor unit appear to contain immunohistochemically identical myosin.

Muscle fibres are not static structures, but easily adapt to altered functional demands, hormonal signals, and changes in neural input. Their dynamic nature makes it difficult to categorise them into distinct units. Therefore, applying any type of rigid classification system may result in an oversimplification which does not take into account the plasticity of the phenotypic expression. Still, fibre typing is an extremely useful tool in many biological fields such as meat science, and on the basis of histochemical stainings, classification of the fibre types help to define functional and metabolic properties of muscle tissue. It must be realised that some properties may change without affecting others or without changing histochemical appearance of a given fibre such as species specificity, developmental or adaptive processes or pathological conditions, and that transient fibre types exist. Therefore, a distinction between specific types must strictly refer to the method that has been used for the typing. Theoretically, there will be at least as many fibre types as there are motor units in a muscle, and within a muscle a fibre type may show a continuum of structural and functional properties which overlap with other fibre types.

Histochemical and biochemical properties of a muscle, such as fibre type, fibre area, oxidative and glycolytic capacity, and glycogen and lipid contents, are factors that may have an influence on meat quality. An important factor for *post mortem* changes and meat quality is the metabolic response that takes place in the different fibre types *pre-slaughter*. Selection for leaner pigs and for a higher proportion of large muscle fibres, especially of type IIB, can result in poor capillarisation and consequently an insufficient delivery of oxygen and substrates and elimination of end products, such as CO_2 and lactate.

In the future, fibre types in pig muscle will probably be investigated with more advanced sensitive techniques where it is possible to look at adaptations in different contractile, sarcoplasmic proteins as well as other muscle proteins that are of importance for cell differentiation and muscle cell metabolism. Furthermore, single fibre dissection and quantitative biochemical analyses may increase the knowledge about metabolic and contractile properties of the muscle fibre.

In general, literature results indicate that it should be possible to include muscle fibre characteristics in breeding schemes for improved meat quality, while preserving optimal production traits. In order to use muscle fibre characteristics in a beneficial way for future breeding programmes, further investigations are needed to better understand the physiological mechanisms and to explain the sometimes controversial results found in different selection studies. This also requires more studies defining predictor muscles, optimal sampling locations, and more rapid and less expensive methods of muscle fibre type determination. Selection experiments based on biochemical and histochemical characteristics determined in biopsies or otherwise and study of the correlated selection responses, may possibly provide better tools to study these relationships.

Acknowledgement

The authors wish to express their gratitude to Dr. Niels Oksbjerg, Dr. Poul Henckel and Dr. José Adalberto Fernández at the Danish Institute of Agricultural Sciences for fruitful discussions and his critical comments during the preparation of this paper. Mrs Marianne Ridderberg is acknowledged for excellent secretarial assistance and linguistic revision of the paper.

References

Armstrong, R.D., P.D. Goldnick, & C.D. Ianuzzo, 1975. Histochemical properties of skeletal muscle fibres in streptozotocin diabetic rats. Tissue Res. 162: 387

Ashmore, C.R. & L. Doerr, 1971. Comparative aspects of muscle fiber types in different species. Exper. Neurol. 31: 408

Bader, R., 1982. Enzymhistochemische und histometrische Untersuchungen an Skelett-muskeln von ausgemästeten, gesunden Schweinen der deutschen Landrasse. Zbl. Vet. Med. A., 29: 443

Bader, R. 1983. Vergleichende histometrische und histologische Untersuchungen an der Skelettmusculatur von Wild- und Hausschweinen. Berlin, München, Tierärtl. Wochenschr. 96: 89

Bar, A. & D. Pette, 1988. Three fast myosin heavy chains in adult rat skeletal muscle. FEBS Lett. 253: 153

Barton-Gade, P., 1981. The measurement of meat quality in pigs post mortem. In: Porcine stress and meat quality-causes and possible solutions to the problems. T. Froysten, E. Slinde, and N. Standal (editors). Agricultural Food Research Society. Ås, Norway. 359 p

Beecher, G.R., R.G.Cassens, W.G. Hoekstra & E.J. Briskey, 1965. Red and white fibre content and associated post-mortem properties of seven porcine muscles. J. Food Sci. 30, 969

Beecher, G.R., L.L.Kastenschmidt, W.G, Hoekstra, R.G. Cassens, & E.J. Briskey, 1969. Energy metabolites in red and white strained muscles of the pig. Agr. Food Chem. 17: 29

Bendall, J.R., 1973. Post mortem changes in muscle. In: Structure and function of muscle. G.H. Bourne (editors), Academic Press, New York, 243-309pp.

Bendall, J.R. & J. Wismed-Pedersen, 1962. Some properties of the fibrillar proteins of normal and watery pork muscle. J. Food Sci., 27: 144

Billeter. R., H. Weber, H. Lutz, Howald, H.M. Eppenberger, & E. Jenny, 1980. Myosin-types in human skeletal muscle fibres. Histochemistry, 65: 249

Briand, M., A.Talmant, Y. Briand, G. Monin & R. Durand, 1981. Metabolic types of muscle in the sheep. 1-Myosin ATPase, glycolytic, and mitochondrial enzyme activities. Eur. J. Appl. Physiol., 46: 347

Brocks, L., R. Klont, W. Buist, K. de Greef, M. Tieman & B. Engel, 1999. The effects of selection of pigs on growth rate vs leanness on histochemical characteristics of different muscles. Submitted.

Brooke, M.H. & K. Kaiser, 1970. Muscle fibre types: How many and what kind? Arch. Neurol., 23: 369

Burke, R.E., D.N. Levine, P. Tsairis & F.E. Zajac, 1973. Physiological types and histo-chemical profiles in motor units of the cat gastrocnemius. J. Physiol. Lond., 234: 723

Cameron, N.D., 1990. Genetic and phenotypic parameters for carcass traits, meat and eating quality traits in pigs. Livest. Prod. Sci. 26: 119

Cameron, N.D., N. Oksbjerg & P. Henckel, 1997. Muscle fibre characteristics of pigs selected for components of efficient lean growth rate. BSAS Annual Meeting, Scarborough, paper No. 32

Cassens, R.G., C.C. Cooper, W.G. Moody, & E.J. Briskey, 1968. Histochemical differentiation of fibre types in developing porcine muscle. J. Anim. Morphol. 15: 135

Cassens, R.G. & C.C. Cooper, 1971. Red and white muscle. Advan. Food. Res., 19: 1

Chasiotis, J., K. Sahlin & E. Hultman, 1983. Regulation of glycogenolysis in human muscle in response to epinephrine infusion. J. Appl. Physiol. 54: 45

Chasiotis, J., 1985. Effect of adrenaline infusion on cAMP and glycogen phosphorylase in fast-twitch and slow-twitch rat muscles. Acta Physiol. Scand. 125: 537.

Chrystall, B.B, S.E. Zobrisky & M.E. Baily, 1969. Longissimus muscle growth in swine. Growth, 33: 361

Cooper, C.C., R.G. Cassens, L.L. Kastenschmidt & E.J. Briskey, 1970. Histochemical characterisation of muscle differentiation. Develop. Biol. 23: 169

Cooper, C.C., R.G. Cassens, L.L. Kastenschmidt & E.J. Briskey, 1971. Activity of some enzymes in developing muscle of the pig. Pediat. Res. 5: 281

Chrystall, B.B., S.E. Zobrisky & M.E. Baily, 1969. Longissimus muscle growth in swine. Growth, 33: 361

Dalrymple, R.H., L.L. Kastenschmidt & R.G. Cassens, 1973. Glycogen and phophorulase in developing red and white muscle. Growth, 37: 19

Davies, A.S., 1972. Postnatal changes in the histochemical fibre types of porcine skeletal muscle. J. Anat, 113: 213

De Vries, A.G., P.G. van der Wal, T. Long, G. Eikelenboom & J.W.M. Merks, 1994. Genetic parameters of pork quality and production traits in Yorkshire populations. Livest. Prod. Sci. 40: 277

De Vries, A.G., A. Sosnicki, J.P. Garnier & G.S. Plastow, 1998. The role of major genes and DNA technology in selection for meat quality in pigs. 44th Int. Congr. Meat Sci. Technol., Barcelona, Spain, 1: 66.

Dietl, G., E. Groeneveld & I. Fiedler, 1993. Genetic parameters of muscle structure in pigs. 44th Annu. Meeting Eur. Assoc. Anim. Prod., Aarhus, Denmark

Engel, W.K., 1962. The essentiality of histo- and cytochemical studies of skeletal muscle in investigation of neuromuscular disease. Neurology, 12: 778

Essen, B., E. Jansson, J. Henriksson, A.W. Taylor & B. Saltin, 1975. Metabolic characteristics of fibre types in human skeletal muscle. Acta Physiol. Scand. 95: 153

Essén-Gustavsson, B., K. Karlström & K. Lundström, 1992. Muscle fibre characteristics and metabolic response at slaughter in pigs of different halothane genotypes and their relation to meat quality. Meat Sci. 31: 1

Essén-Gustavsson, B. & A. Lindholm, 1984. Fiber types and metabolic characteristics in muscles of wild boars, normal and halothane sensitive Swedish Landrace pigs. Comp. Biochem. Physiol. 78A: 67

Essén-Gustavsson, B. & S. Fjelkner-Modig, 1985. Skeletal muscle characteristics in different breeds of pigs in relation to sensory properties of meat. Meat Sci. 13: 33

Fernandez, X. & E. Tornberg, 1991. A review of the causes of variation in muscle glycogen content and ultimate pH in pigs. J. Muscle Foods, 2: 209

Fernandez, X., M.C. Meunier-Salaün & P. Mormède, 1994a. Agonistic behaviour, plasma stress hormones, and metabolites in response to dyadic encounters in domestic pigs: interrelationships and effect of dominance status. Physiol. Behav. 56: 841.

Fernandez, X., M.C. Meunier-Salaün & P. Ecolan, 1994b. Glycogen depletion according to muscle and fibre types in response to dyadic encounters in pigs relationships with plasma epinephrine and aggressive behaviour. Comp. Biochem. Physiol. 109A: 869

Fernandez, X., P. Levasseur, P. Ecolan & W. Wittmann, 1995a. Effect of epinephrine administration on glycogen metabolism in red and white muscle of anaesthetized pigs (sus scrofa domesticus). J. Sci. Food Agric. 68: 231

Fernandez, X., M.C. Meunier-Salaün, P. Ecolan & P. Mormède, 1995b. Interactive effect of food deprivation and agonistic behavior on blood parameters and muscle glycogen in pigs. Physiol. Behav. 58: 337

Fiedler, I., J. Wegner & K.D. Feige, 1991. Ein Wachstumsmodell für den Faserdurchmesser in zwei Muskeln des Schweines. Arch. für Tierzucht, 34: 57.

Garnett, R.A.F., M.J. O'Donovan, J.A. Stephens & A. Taylor, 1978. Motor unit organisation of human gastrocnemius. J. Physiol. 287: 33

Gauthier, G.F., R.E. Burke, S. Lowry & A.W. Hobbs, 1983. Myosin isozymes in normal and cross-reinnervated cat skeletal muscle fibres. J. Cell Biol. 97: 756

Goodyear, L.J., M.F. Hirschman, R.J. Smith, & E.S. Horton, 1991. Glucose transporter number, activity and isoform content in plasma membranes of red and white skeletal muscle. Am. J. Physiol. 261 (Endocrinol Metab. 24): E556

Gorski, J., 1978. Glycogenolytic effect of adrenaline in skeletal muscles of rats adapted to endurance exercise. Acta Physiol. Pol. 29: 5

Gorza, L., 1990. Identification of a novel type 2 fibre population in mammalian skeletal muscle by combined use of histochemical myosin ATPase and anti-myosin monclonal antibodies. J. Histochem. Cytochem. 38: 257

Greenhaff, P.L., J.M. Ren, K. Soderlund & E. Hultman, 1991. Energy metabolism in single human muscle fibres during contraction without and with epinephrine infusion. Am. J. Physiol. 260 (Endocrinol. Metab. 23): E713

Hatton, M.W.C., R.A. Lawrie, P.W. Ratcliff & N. Wayne, 1972. Effects of preslaugther adrenaline injection on muscle metabolites and meat quality of pigs. J. Food Technol. 7: 443

Hedrick, H.B., F.C. Parrish & M.E. Bailey, 1964. Effect of adrenaline stress on pork quality. J. Anim. Sci. 23: 225

Hegarty, P.V.J., L.C. Gundlach & C.E. Allen, 1973. Comparative growth of porcine skeletal muscle using an indirect prediction of muscle fibre number. Growth, 37: 333

Henckel, P., N. Oksbjerg, E. Erlandsen, P. Barton-Gade & C. Bejerholm, 1997. Histo- and biochemical characteristics of the Longissimus dorsi muscle in pigs and their relationships to performance and meat quality. Meat Sci. 47: 311

Hintz, C.S., E.F. Coyle, K.K. Kaiser M.M.Y. Chi & O.H. Lowry, 1985. Comparison of muscle fiber typing by quantitative enzyme assays and by myosin ATPase staining. J. Histochem. Cytochem. 32: 655

Hocquette, J.F., I. Ortigues, D. Pethick, P. Herpin & X. Fernandez, 1998. Nutritional and hormonal regulation of energy metabolism in skeletal muscles of meat-producing animals. Livest. Prod. Sci. 56: 115

James, D.E., E.W. Kraegen & D.J. Chisholm, 1985. Muscle glucose metabolism in exercising rats: comparison with insulin stimulation. Am. J. Physiol. 248 (Endocrinel Metab. 11): E-575.

Jensen, J., H.A. Dahl & P.K. Opstad, 1989. Adrenaline mediated glycogenolysis in different skeletal muscle fibre types in the anesthetized rat. Acta Physiol. Scand. 136: 229

Jul, M., P. Zeuthen, 1981. Quality of pig meat for fresh consumption. Progress in food and nutrition science. Pergamon Press, Oxford., 4: 6.iet, 350, Copenhagen

Karlsson, A., 1993. Porcine muscle fibres –biochemical and histochemical properties in relation to meat quality. Swedish University of Agricultural Sciences, Department of Food Sci., Uppsala, Sweden, Ph-D Thesis.

Karlsson, A., B. Essén-Gustavsson & K. Lundström, 1992. Muscle glycogen depletion pattern in Longissimus dorsi muscle of pigs' fed high- and low-protein diet. 38th International Congress of Meat Science and Technology. Clermont Ferrand, France. 375-8

Karlsson, A., A.-C. Enfält, B. Essén-Gustavsson, K. Lundström, L. Rydhmer & S. Stern, 1993. Muscle histochemical and biochemical properties in relation to meat quality during selection for increased lean tissue growth rate in pigs. J. Anim. Sci. 71: 930

Karlström, K., B. Essén-Gustavsson, M. Jensen-Waern & A. Lindholm, 1995. Fibre distribution, capillarization and enzymatic profile of locomotor and nonlocomotor muscles of wild boar and domestic pig. In: Swedish University of Agricultural Sciences, Department of Medicine and Surgery, Sweden, Ph-D Thesis by Karlström, K

Kelso, T.B., D.R. Hodgson, A.R. Visscher & P.D. Gollnick, 1987. Some properties of different skeletal muscle fibre types: comparison of reference bases. J. Appl. Physiol. 62: 1436

Kennedy, J.M., F.S.F. Mong, & J.L. Poland, 1986. Histochemical fiber type distribution and epinephrine sensitivity in normal and cross-transplanted rat muscles. Can J. Physiol. Pharmacol. 65: 1205

Kiessling, K-H., K. Lundström, H. Petersson & H. Stalhammar, 1982. Age and feed related to changes of fiber composition in pig muscle. Swedish J. Agric. Res. 12: 69

Kiessling, K-H. & I. Hansson, 1983. Fiber composition and enzyme activities in pig muscles. Swedish J. Agric. Res. 13: 257.

Klont, R.E., ELambooy & J.G. van Logtestijn, 1993. Effect of preslaughter anaesthesia

Kuchel, O., 1991. Stress and catecholamines. Methods Achieve Exp. Pathol. 14: 80

Kugelberg, E., 1976. Adaptive transformation of rat soleus motor units during growth. J. Neurol. Sci. 27: 269

Laborde, D., A. Talmant & G. Monin, 1985. Activités enzymatiques et contractiles de 30 muscles de porc: Relation avec le pH ultime atteint après la mort. Reprod. Nutr. Dev. 25: 619

Lacourt, A. & P.V. Tarrant, 1985. Glycogen depletion patterns in myofibres of cattle during stress. Meat Sci. 15: 85.

Larzul, C., L. Lefaucheur, P. Ecolan, J. Gogué, A. Talmant, P. Sellier, P. Le Roy, & G. Monin, 1997. Phenotypic and genetic parameters for longissimus muscle fiber characteristics in relation to growth, carcass, and meat quality traits in Large White pigs. J. Anim. Sci. 75: 3126

Lefaucheur, L. & P. Vigneron, 1986. Post-natal changes in some histochemical and enzymatic characteristics of three pig muscles. Meat Sci. 16: 199.

Leseigneur-Meyneir, A. & G. Gandemer, 1991. Lipid composition of pork muscle in relation to the metabolic type of the fibre. Meat Science, 29: 229

Lind, A. & D. Kernell, 1991. Myofibrillar ATPase histochemistry of rat skeletal muscles: A "two-dimensional" quantitative approach. J. Histochem Cytochem. 39: 589

Lopez-Bote, C., P.D. Warriss & S.N. Brown, 1989. The use of muscle protein solubility measurements to assess pig lean meat quality. Meat Science, 26: 167

Lowry, O.H., C.V. Lowry, M.M.-Y. Chi, C.S. Hintz & S. Felder, 1980. Enzymological heterogeneity of human muscle fibres. In: Plasticity of muscle. D. Pette (editors), de Gruyter, Berlin, 3-18

Lundström, K., B. Essén-Gustavsson, M. Rundgren, I. Edfors-Lilja & G. Malmfors, 1989. Effect of halothane genotype on muscle metabolism at slaughter and its relationship with meat quality: A within-litter comparison. Meat Sci. 25: 251

Lutz, H., E. Weber, R. Billeter & E. Jenny, 1979. Fast and slow myosine within single skclctal muscle fibres of adult rabbits. Nature, 281: 142

Mauch, A. & J. Marinesco, 1934. Recherches sur la viande de Mangalitza comparée a celle de races York, Lincoln et leurs métis. Ann. Inst. Zootechn. Bukarest, 3: 154

Maltin, C.A., C.C. Warkup, K.R. Matthews, C.M. Grant, A.D. Porter M.I. & Delday, 1997. Pig muscle fibre characteristics as a source of variation in eating quality. Meat Sci. 47: 237

McCampbell, H.C., F.M. Griffin, R.W. Seerley & C.W. Foley, 1974. Wild and domestic swine: growth and composition. J. Animal Sci. 38: 220

Miller, L.R., V.A. Garwood & M.D. Judge, 1975. Factors affecting porcine muscle fiber type, diameter and number. J. Anim. Sci. 41: 66

Monin, G. & P. Sellier, 1985. Pork of low technological quality with normal rate of muscle pH fall in the immediate post-mortem period: The case of the Hampshire breed. Meat Sci. 13: 49

Monin, G., A. Talmant, D. Laborde, M. Zabari & P. Sellier, 1986. Compositional and enzymatic characteristics of the longissimus dorsi muscle from Large White, halothane-positive and halothane-negative Pietrain, and Hampshire pigs. Meat Sci. 16: 307

Monin, G., Â. Mejenes-Quijano, A. Talmant & P. Sellier, 1987. Influence of breed and muscle metabolic type on muscle glycolytic potential and meat pH in pigs. Meat Science, 29: 149

Monin, G. & A. Quali, 1991. Muscle differentiation and meat quality. In: Developments in Meat Science – 5. R. Lawrie (editor), Elsevier Applied Sciences, London, 89-159

Moody, W.G., M.B. Enser, J.D. Wood, D.J. Restall & D. Lister, 1978. Comparison of fat and muscle development in Pietrain and Large White piglets. J. Anim. Sci. 46: 618

Nemeth, P.M. & R.S. Wilkinson, 1990. Metabolic uniformity of the motor unit. In: The dynamic state of muscle fibers. D. Pette (editor), de Gruyter, Berlin, 233-245

Novikoff, A.B., W.-Y. Shin & J. Drucker, 1961. Mitochondrial localization of oxidative enzymes: Staining results with two tetrazolium salts. Biophys. Biochem. Cytol. 9: 47

Nwoye, L., W.F.H.M. Mommaerts, D.R. Simpson, K. Seraydarian & M. Marusich, 1982. Evidence for a direct action of thyroid hormone in specifying muscle properties. Am. J. Physiol. 242: R401

Ogata, T., 1958. A histochemical study of the red and white muscle fibers, I: Activity of the succinoxydase system in the muscle fibers. Acta Med. Okayama, 12: 216

Oksbjerg, N., A. Blackshaw, P. Henckel, J.A. Fernández & N. Agergaard, 1990. Alterations in protein accretion and histochemical characteristics of the M. longissimus dorsi in pigs caused by Salbutamol (a β-adrenergic agonist). Acta Agric. Scand. 40: 397

Oksbjerg, N., P. Henckel & T. Rolph 1994a. Effects of salbutamol, a β_2-adrenergic agonist, on muscles of growing pigs fed different levels of dietary protein, I. Muscle fibre properties and muscle protein accretion. Acta Agric. Scand., Sect. A, Animal Sci. 44: 12

Oksbjerg, N., P. Henckel, T. Rolph & E. Erlandsen, 1994b. Effects of salbutamol, a β_2-adrenergic agonist, on muscles of growing pigs fed different levels of dietary protein, II. Aerobic and glycolytic capacities and glycogen metabolism of the longissimus dorsi muscle. Acta Agric. Scand., Sect. A, Animal Sci. 44: 20

Oksbjerg N. & M.T. Sørensen, 1996. Effects of ephedrine and caffeine on chemical composition and histochemistry of muscles in the pig. Acta Agric. Scand. Sect. A, Animal Sci. 46: 125

Peter, J.B., R.J. Barnard, V.R. Edgerton, C.A. Gillespie & C.E. Stempel 1972. Metabolic profiles of three fiber types of skeletal muscle in guinea pigs and rabbits. Biochemistry, 11: 2627

Peterson, J.S., P. Henckel, N. Oksbjerg & M.T. Sørensen, 1998. Adaptions in muscle fibre characteristics induced by physical activity in pigs. Animal Science, 66: 733

Petersen, J.S., P. Henckel & S. Stoier, 1997. Muscle physiological traits and meat quality in Danish Landrace Anno, 1976 and Anno1995. 48th Annu. Meet. Assoc. Anim. Prod., Vienna, Austria

Pette, D. & R.S. Staron, 1990. Cellular and molecular diversity of mammalian skeletal muscle fibers. Rev. Physiol. Biochem. Pharmacol. 116: 1

Pette, D. & G. Vrbodá, 1985. Invited review: neural control of phenotypic expression in mammalian muscle fibers. Muscle Nerve, 8: 676

Pierobon-Bormioli, S., S. Sartore, L. Dalla Libera, M. Vitadello & S. Schiaffino, 1981. Fast isomyosins and fibre types in mammalian skeletal muscle. J. Histochem. Cytochem. 29: 1179

Rahelic, S. & S. Puac, 1980-81. Fibre types in longissimus dorsi from wild and highly selected pig breeds. Meat Sci. 5: 439

Rao, M.V. & N.F.S. Gault, 1979. The influence of fibre type composition and associated biochemical characteristics on the acid buffering capacity of several beef muscles. Meat Sci. 26: 5

Rede, R., V. Pribisch & S. Rahelic, 1986. Untersuchungen über die Beschaffenheit von Schlacttierkörpern und Fleisch primitiver und hochselektierter Schweinerassen. Fleischwirtsch. 66: 898

Renou, J.P., P. Canioni, P. Gatellier, C. Valin & P.J. Cozzone, 1986. Phosphorus 31 nuclear magnetic resonance study of post mortem catabolism and intracellular pH in intact excised rabbit muscle. Biochimie, 68: 543

Richter, E.A., N.B. Ruderman, H. Gavras, E.R. Belur & H. Galbo, 1982. Muscle glycogenolysis during exercise: dual control by epinephrine and contractions. Am. J. Physiol. 242 (Endocrinol. Metab. 5), E25-E32

Robbins, N., G. Karpati & W.K. Engel, 1969. Histochemical and contractile properties in the cross-innerveated guinea pig soleus muscle. Arch. Neurol. 20: 318

Rubli, H., 1931. Die Myologi des Wildschweines. Arch. Sozialantrop. Rassenhyg. 5: 391

Ruusunen, M., 1989. Korrelationen mellan vattenhållande fürmåga och muskelfibersamman-sättning i olika typer av svinmuskler. Köttkvalitet hos Våra Slaktdjur, NJF-seminarium Nr 183. pp. 33-40. Swedish University of Agricultural Sciences Department of Food Science, K. Lundström and G. Malmfors (editors). Uppsala, 262

Sair, R.A., L.L. Kastenschmidt, R.G. Cassens & E.J. Briskey, 1972. Metabolism and histo-chemistry of skeletal muscle from stress-susceptible pigs. J. Food Sci. 37: 659

Salomon, F.-V., G. Michel & F. Gruschwitz, 1983. Zur Entwicklung von Fasertypenkompo-sition und Faserdurchmesser im M. longissimus des Hausschweines. Anat. Anz., Jena 154: 69

Schiaffion, S., E. Ausoni, L. Gorza, L. Saggin, K. Gundersen & T. Lømo, 1988. Myosin heavy chain isoforms and velocity of shortening of type 2 skeletal muscle fibres. Acta Physiol. Scand. 134: 575

Schiaffion, S., L. Gorza, S. Sartore, L. Saggin, S. Ausoni, M. Vianello, K. Gunderson & T. Lømo, T., 1989. Three myosin heavy chain isoforms i type 2 skeletal fibres. J. Muscle Res. Cell Motil. 10: 197

Schlegel, O., 1982. Untersuchungen über die Fasertypenverteilung und Fasertypenquer-schnittsflächen im M. longissimus dorsi und M. semitendinosus von trainierten Haus- und immobil gehaltenen Wildschweinen . Hannover, Tierärztliche Hoch-schule, Physiologisches Institut. Diss. 99 p

Sellier, P., 1988. Aspects génétiques des qualités technologiques et organoleptiques de la viande cheze le porc. Journé Rechn. Porc. En France, 20: 227

Severini, M., G. Cenci & A. Vizzani, 1989. Post mortem glycogenolysis and pigmeat quality. In Proceedings, 35th ICoMST, Aug. 20-25, 1989. Roskilde, Denmark, Vol. III, 1137

Solomon, M.B., R. G. Campbell, N.C. Steele, 1990. Effect of sex, exogenous porcine somatotropin on longissimus muscle fiber characteristics of growing pigs. J. Anim. Sci. 68: 1176

Somers, C. & P.V. Tarrant, 1984. Evaluation of some objective methods for measuring pork quality. Proc. Scient. Meeting biophysical PSE-muscle analysis, Vienna 230

Sonesson, A.K., K.H. De Greef & T.H.E. Meuwissen, 1998. Genetic parameters, trends of meat quality and carcass composition and performance traits in two selected lines of large white pigs. Livest. Prod. Sci. Accepted for publication

Sørensen, M.T., N. Oksbjerg, N. Agergaard & J.S. Petersen, 1996. Tissue deposition rates in relation to muscle fibre and fat cell characteristics in lean female pigs (Sus scrofa) following treatment with porcine growth hormone (pGH. Comp. Biochem. Physiol, 113A: 91

Staun, H., 1963. Various factors affecting number, size of muscle fibers in pig. Acta Agriculture Scandinavica, 13: 293

Staron, R.A. & D. Pette, 1987. The multiplicity of myosin light and heavy chain combinations in histochemically typed single fibres. Biochem. J. 243: 687

Suzuki, A. & R.G. Cassens, 1980. A histochemical study of myofibre types in muscle of the growing pig. J. Anim. Sci. 51: 1449

Suzuki, A., N. Kojima, Y. Ikeuchi, S. Ikarashi, N. Moriyama, T. Ishizuka & H. Tokushige, 1991. Carcass composition and meat quality of Chinese purebred and European x Chinese crossbred pigs. Meat Sci. 29: 31

Swatland, H.J., 1976. Longitudinal fascicular growth in the porcine longissimus muscle. J. Anim. Sci. 42: 63

Swatland, H.J., 1978. A note on the physiological development of a red muscle from the ham region of the pork carcass. Animal Production 27: 229

Szentkuti, L. & R.G. Cassens, 1978. Die Verteilung der Fasertypen I, IIA, IIB im M. longissimus dorsi und M. semitendinosus von Schweinen verschiedenen Alters. Dtsch. Tierärztl. Wochenschr. 85: 23

Szentkuti, L., B. Niemeyer & O. Schlegel, 1981. Vergleichende Untersuchung von Muskelfasertypen mit der Myosin-ATPase Raktion im M. longissimus dorsi von Haus- und Wildschweinen. Dtsch. Tierärztl. Wschr. 88: 407

Szentkuti, L. & O. Schlegel, 1985. Genetische und funktionelle Einflüsse auf Fasertypen-anteile und Faserdurchmesser im M. longissimus dorsi und M. semitendinosus von Schweinen. Untersuchungen an trainierten Haus- und immobil Gehaltenen Wildschweinen. Dtsch. Tierärztl. Wschr. 92: 93

Termin, A., R.S. Staron & D. Pette, 1989. Myosin heavy chain isoforms in histochemically defined fiber types of rat muscles. Histochem. 92: 453

Touraille, C., G. Monin & C. Legault, 1989. Eating quality of meat from European x Chinese crossbred pigs. Meat Sci., 25: 177

Van den Hende, C., E. Muylle, W. Oyaert & P. De Roose, 1972. Changes in muscle characteristics in growing pigs: Histochemical and electron microscopie study. Zbl. Vet. Med., A19: 102

Villa-Moruzzi, E., E. Bergamini & Z.G. Bergamini, 1981. Glycogen metabolism and the function of fast and slow muscles of the rat. Pflügers Archiv 391: 338

Von Lengerken, G., S. Maak, M. Wicke, I. Fiedler & K. Ender, 1994. Suitability of structural and functional traits of skeletal muscle for genetic improvement of meat quality in pigs. Arch. Tierz. Dummerstorf, 37: 133

Wegner , J. & K. Ender, 1990. Mikrostrukturelle Grundlagen des Wachstums von Muskel- und Fettgewebe und die Beziehung zu Fleischansatz und Fleischbeschaffenheit. Fleischwirtschaft, 70: 337

Weiler, U., H.-J. Appell, M. Kermser, S. Hofäcker & R. Claus, 1995. Consequences of selection on muscle composition. A comparative study on Gracilis muscle in wild and domestic pigs. Anat. Histol. Embryol. 24: 77

Wismer-Pedersen, J., 1959. Quality of pork in relation to rate of pH change post-mortem. Food Res. 24: 711

Selection progress of intramuscular fat in Swiss pig production

D. Schwörer[1], A. Hofer[2], D. Lorenz[1] & A. Rebsamen[1]

[1] *Swiss Pig Performance Testing Station, 6204 Sempach, Switzerland*
[2] *Institute of Animal Science, Swiss Federal Institute of Technology (ETH), CLU, 8092 Zürich, Switzerland*

Summary

The Swiss Pig Performance Testing Station started in the year 1980 with the evaluation of intramuscular fat (IMF: amount of free fat in the 10th rib section of the *M. longissimus dorsi*). Since the year 1985 all full sibs of the Testing Station were routinely analysed on IMF by Near Infrared Reflection. Until the end of 1998 a total of 57975 animals were tested. Inclusion of IMF in the combined selection index began in the year 1989 (desired selection progress per generation: 0.03 % IMF). Since the year 1997 IMF is also considered in the BLUP evaluation system.

Selection for higher IMF has proved to be effective. In Swiss Large White (SLW) IMF increased phenotypically from 1.0 % in the year 1987 to 1.9 % in the year 1998, and in Swiss Landrace (SL) from 0.8 % to 1.6 %. The standard deviation remained constant. The BLUP-breeding values were used for the estimation of genetic trends. In the years 1991-1998 IMF increased in SLW genetically by 0.3 % and in SL by 0.2 %. In the same period a genetic increase in premium cuts of 1.3 % in SLW, of 1.7 % in SL and in daily gain of 18 g in SLW, of 25 g in SL were estimated. The genetic improvement in feed conversion was in both breeds 0.1 kg/kg.

Keywords: Intramuscular fat, Pigs, Selection progress, Swiss Pig Performance Testing Station

Introduction

Considerable progress was realised in Swiss pig breeding for fattening and slaughtering traits in the last decades. Steps were necessary to prevent a deterioration of meat and fat quality. Research on meat and fat quality at the Swiss Pig Performance Testing Station (MLP) led to the following breeding measures concerning intramuscular fat (IMF):

1980: Inclusion of IMF in breeding considerations, evaluation of the methods and estimation of the phenotypic and genetic parameters (Schwörer *et al.*, 1986; Schwörer *et al.*, 1987; Schwörer *et al.*, 1990; Hofer & Schwörer, 1995).

1985: Start of the routine evaluation of IMF by Near Infrared Reflection.

1989: Inclusion of IMF in the combined selection index (Morel *et al.*, 1988).

1997: Estimation of the breeding value by BLUP, including premium cuts, daily gain, feed conversion, meat quality (pH1, reflection) and IMF (Hofer *et al.*, 1996).

Objectives

The inclusion of IMF in the combined selection index and the BLUP evaluation system should not only improve the sensory quality, but should also indirectly stop the reduction in carcass fatness and improve the fat quality of the fatty tissues; resulting in more fat but less water and polyenic fatty acids in the fatty tissues (Schwörer et al., 1995).

Methods

Full sib animals (females, castrates) at the Swiss Pig Performance Testing Station are kept in groups of 4 or 10 animals (fattening period: 30-103 kg liveweight). IMF is measured as the percentage of free fat in the 10th rib section of the M. longissimus dorsi by Near Infrared Reflection (InfraAlyzer 450, Bran+Luebbe). The Soxleth fat extraction (Trichloraethane), without previous acid hydrolysis (Soxtec HT System, Tecator) served as a reference method. Multilinear regression showed the following calibration results: $r = 0.92$, SE $= 0.25$ (Schwörer, 1992). BLUP breeding values were used for the estimation of genetic trends according to Hofer et al., 1996.

Results

Selection on IMF proves to be successful (Figure 1). The decline in IMF stopped. The amount of IMF has not yet reached the optimum of 2.0 % free fat in our breeds. In Swiss Large White (SLW) IMF increased phenotypically from 1.0 % in the year 1987 to 1.9 % in the year 1998, and in Swiss Landrace (SL) from 0.8 % to 1.6 %. The standard deviation remained constant during this period (Figure 2).

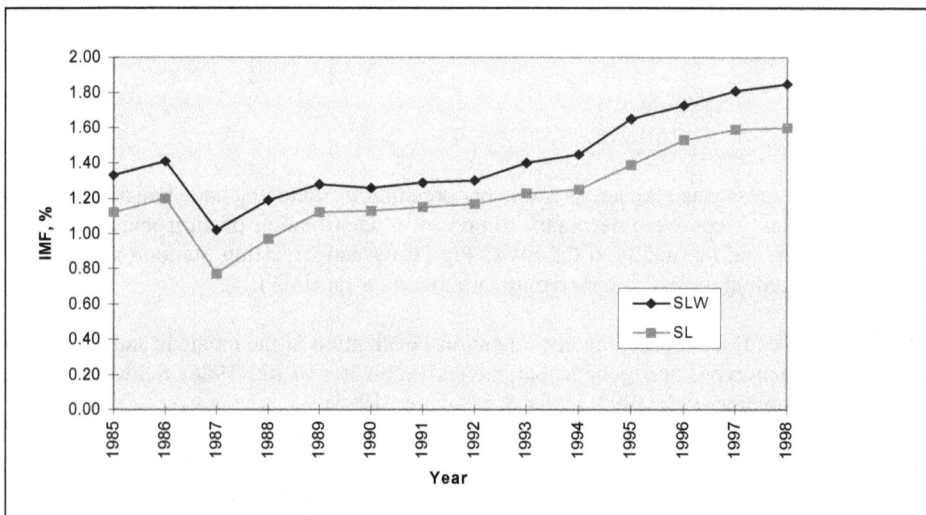

Figure 1. Content of intramuscular fat (IMF, %) in Swiss Large White (SLW) and Swiss Landrace (SL) pigs (years 1985-1998, full-sib test).

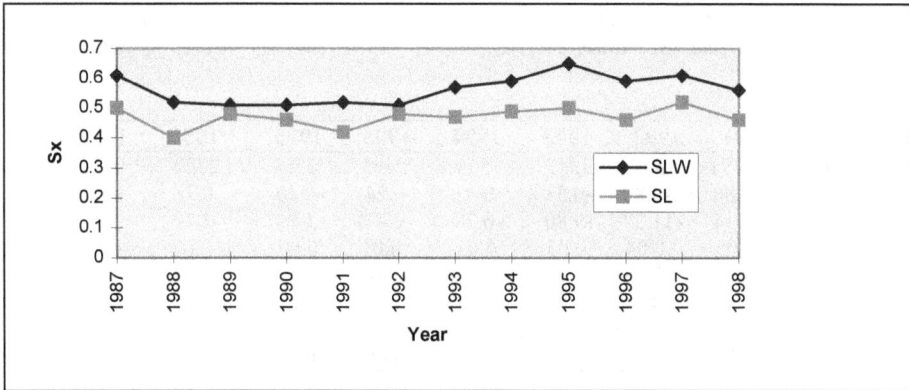

Figure 2. Standard deviation of intramuscular fat (sx) in Swiss Large White (SLW) and Swiss Landrace (SL) pigs (years 1987-1998, full-sib test).

The percentage of animals with more than 2 % of IMF in the loin is increasing (Figure 3). In SLW it increased from 8.1 % in the year 1990 to 32.1 % in the year 1998, and in SL from 4.1 % to 15.6 %. In the same period the percentage of animals with less than 1 % of IMF decreased drastically (SLW: from 34.2 % to 1.6 %; SL from 45.5 % to 3.3 %).

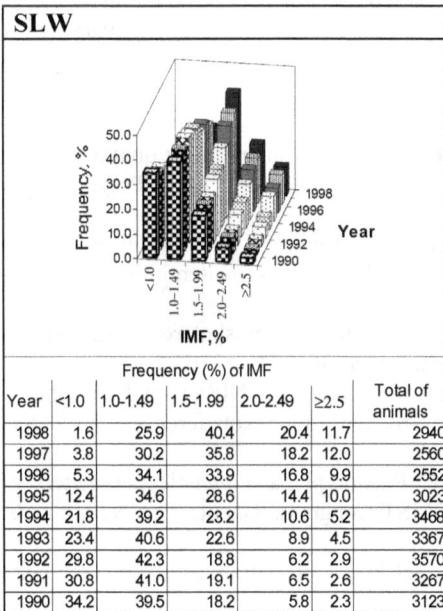

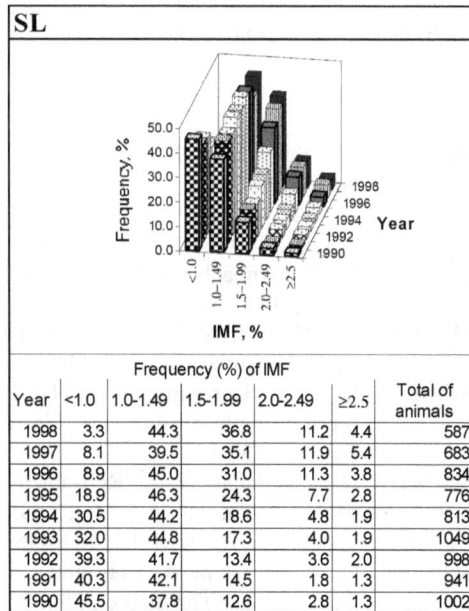

SLW

Frequency (%) of IMF

Year	<1.0	1.0-1.49	1.5-1.99	2.0-2.49	≥2.5	Total of animals
1998	1.6	25.9	40.4	20.4	11.7	2940
1997	3.8	30.2	35.8	18.2	12.0	2560
1996	5.3	34.1	33.9	16.8	9.9	2552
1995	12.4	34.6	28.6	14.4	10.0	3023
1994	21.8	39.2	23.2	10.6	5.2	3468
1993	23.4	40.6	22.6	8.9	4.5	3367
1992	29.8	42.3	18.8	6.2	2.9	3570
1991	30.8	41.0	19.1	6.5	2.6	3267
1990	34.2	39.5	18.2	5.8	2.3	3123

SL

Frequency (%) of IMF

Year	<1.0	1.0-1.49	1.5-1.99	2.0-2.49	≥2.5	Total of animals
1998	3.3	44.3	36.8	11.2	4.4	587
1997	8.1	39.5	35.1	11.9	5.4	683
1996	8.9	45.0	31.0	11.3	3.8	834
1995	18.9	46.3	24.3	7.7	2.8	776
1994	30.5	44.2	18.6	4.8	1.9	813
1993	32.0	44.8	17.3	4.0	1.9	1049
1992	39.3	41.7	13.4	3.6	2.0	998
1991	40.3	42.1	14.5	1.8	1.3	941
1990	45.5	37.8	12.6	2.8	1.3	1002

Figure 3. Frequency of IMF classes of full sib tested Swiss Large White (SLW) and Swiss Landrace (SL) pigs (years 1990-1998).

The BLUP-breeding values were used for the estimation of genetic trends. In the years 1991-1998 IMF increased in SLW genetically by 0.3 % and in SL by 0.2 %. In the same period a genetic increase in premium cuts of 1.3 % in SLW, of 1.7 % in SL and in daily gain of 18 g in SLW, of 25 g in SL were estimated. The genetic improvement in feed conversion was in both breeds 0.1 kg/kg (Table 1).

71

Table 1. *Estimation of genetic trends for full sib tested Swiss Large White (SLW) and Swiss Landrace (SL) animals (years 1991-1998) (average breeding values/ birth year).*

Birth year	1991	1992	1993	1994	1995	1996	1997	1998
Animals (SLW)	1391	3573	3247	3554	2495	2564	2722	1794
PC	-0.88	-0.89	-0.53	-0.41	-0.24	-0.10	0.18	0.39
DG	-12.14	-11.37	-9.80	-6.78	-2.47	-1.94	2.42	5.86
FC	0.05	0.04	0.03	0.03	0.01	0.00	-0.02	-0.03
IMF	**-0.19**	**-0.16**	**-0.12**	**-0.07**	**-0.02**	**0.03**	**0.05**	**0.08**
pH1	-0.01	-0.01	-0.01	-0.01	0.00	0.01	0.01	0.01
pH30	0.00	0.00	0.00	0.00	0.00	0.00	0.00	0.00
Reflection	-0.07	-0.09	-0.02	0.02	-0.06	-0.02	0.02	-0.01

Birth year	1991	1992	1993	1994	1995	1996	1997	1998
Animals (SL)	513	915	914	868	814	681	576	431
PC	-1.29	-1.12	-0.40	-0.37	-0.33	-0.31	0.02	0.43
DG	-28.75	-28.09	-17.73	-10.10	-7.14	2.41	-2.10	-4.24
FC	0.10	0.08	0.04	0.03	0.02	0.02	0.01	0.01
IMF	**-0.08**	**-0.05**	**-0.06**	**-0.02**	**0.01**	**0.03**	**0.07**	**0.09**
pH1	-0.03	-0.02	-0.03	-0.02	-0.01	-0.01	0.00	0.00
pH30	0.00	0.00	0.00	0.00	0.00	0.00	0.00	0.00
Reflection	0.33	0.18	0.47	0.45	0.30	0.27	0.09	0.05

References

Hofer, A. & D. Schwörer, 1995. Genetic parameters of production and meat quality traits in station tested Swiss Large White pigs. 46 th Meeting of EAAP, Prague, Paper G.5.15: 1-4

Hofer, A., D. Lorenz, D. Schwörer & A. Rebsamen, 1996. Zuchtfortschritt bei Prüftieren der MLP Sempach. Vortragstagung: Aktuelle Forschungsarbeiten mit Nutztieren, ETH Zürich, Schweiz, I, 1

Morel, P., D. Schwörer & A. Rebsamen, 1988. Einbau des intramuskulären Fettes in den Selektionsindex der MLP. Der Kleinviehzüchter, 36: 1341-1350

Schwörer, D., 1992. Der Einsatz der Nah-Infrarot-Spektroskopie (NIRS) im Fleischlabor. Vortragstagung: Aktuelle Forschungsarbeiten mit Nutztieren, Weinfelden, Schweiz, IV, 1

Schwörer, D., J.K. Blum & A. Rebsamen, 1986. Phenotypic and genetic parameters of intramuscular fat in pigs. 32 nd ICOMST-Congress, Ghent, The Netherlands, Proceeding II, Paper 9:3, 433-436

Schwörer, D., P. Morel & A. Rebsamen, 1987. Selektion auf intramuskuläres Fett beim Schwein. Der Tierzüchter, 39: 392-394

Schwörer, D., P. Morel, A. Prabucki & A. Rebsamen, 1990. Genetic parameters of intramuscular fat content and fatty acids of pork fat. 4th World Congress on Genetics applied to Livestock Production, Edinburgh, Great Britain, Proceeding XV, 545-548

Schwörer, D., A. Rebsamen & D. Lorenz, 1995. Selection on intramuscular fat in Swiss pig breeds and the importance of fatty tissue quality. 2nd Dummerstorf Muscle Workshop: Muscle Growth and Meat Quality, Rostock, Germany, 116-124

Nutritional and genetic influences on meat and fat quality in pigs

K. Nürnberg, G. Kuhn, U. Küchenmeister & K. Ender*

Department of Muscle Biology & Growth, Research Institute for the Biology of Farm Animals, D-18196 Dummerstorf, Germany

Summary

The results of three experiments show the influence of nutrition and genetics on meat and fat quality. N-3 fatty acid enriched diet fed to pigs did not affect the meat quality (Exp. 1), however this diet increased the percentage of n-3 fatty acids in polar fractions of muscle. A study with Saddle Back pigs (SB) and German Landrace pigs (GL) indicated differences in meat and fat quality (Exp. 2). The larger fat content of SB pigs was related to an increased intramuscular fat concentration of *longissimus* muscle. The saturated fatty acid concentration was higher and that of the polyunsaturated fatty acids was significantly lower in SB pigs than in GL muscle. Heterozygotes (NP) of SB showed higher values for drip loss and meat colour compared to homozygotes (NN) SB. It can be concluded that a high fat content is no guarantee for PSE free meat.

A mutation of the calcium release channel (CRC) caused an increased metabolism post mortem resulting in cell injury and inferior meat quality in Pietrain pigs (Exp. 3). Malignant hyperthermia susceptible pigs (MHS, PP) had a higher concentration of n-3 fatty acids in muscle phospholipids, enhanced peroxidation of lipids and an insufficient SR-Ca^{2+} regulation in comparison to stress resistant pigs. The results lead to the recommendation to exclude MHS pigs from breeding programs.

Keywords: Meat quality, Fat quality, Fatty acids, Membrane, Pigs

Material and methods

Experiment 1: 20 castrated pigs [(Pi x Ha) (Large White x GL)] were divided into two groups at an average weight of 40 kg. The control feed contained 1.3 g n-3 fatty acids/kg diet (control group) and the experimental feed contained 14 g n-3 fatty acids/kg diet (n-3 diet group). Fish oil was added daily to the feed mixture of the n-3-diet group as a supplement. Details of housing, feeding, and analytical methods used are described by Nürnberg *et al.* (1998).

Experiment 2: A total of 58 castrated pigs [German Saddle Back (SB), NN n= 19, NP n=9); German Landrace (GL), NN n = 30] were used in two experiments at an average weight of 110 kg at slaughter. Detailed methods for housing and feeding are described by Kuhn *et al.* [1997].

Experiment 3: A total of 26 female Pietrain pigs were used for the study. The animals originated from two different Pietrain lines of the same breeding farm. No mutations in the RYR-1 gene were present in 17 of the pigs (homozygous, MH resistant = MHR) and 9 pigs were homozygous for the mutant allele (MH susceptible = MHS). The pigs were slaughtered at 110 kg live weight. Samples of the *longissimus* muscle were removed from the carcass immediately after electrical stunning and exsanguination (0 h). Subsequent *post mortem* (p.m.) samples were taken at 45 minutes, 4 h, and 22 h after chilling. All samples were immediately homogenised or frozen in liquid nitrogen. For details see Küchenmeister *et al.* (1999a, b).

Results

Experiment 1: There was no influence of dietary inclusion of fish oil fatty acids on muscle meat percentage, backfat thickness, and meat quality traits like pH - value, water binding capacity, and meat colour (Table 1). Otten (1993) also reported no effect of exogenous n-3 fatty acids on carcass composition and meat quality in pigs. The exogenous application of n-3 fatty acids caused a significantly reduced intramuscular fat content in muscle. The intake of dietary n-3 fatty acids increased the percentages of these fatty acids in the phospholipids of skeletal muscle significantly, compared to control (30.7 % vs. 3.9 % n-3 fatty acids), while the n-6 fatty acid concentration was reduced (Figure 1).

Table 1   Meat quality.

| | Control | | n-3 diet | |
	LSM	SEM	LSM	SEM
Longissimus muscle				
Intramuscular fat [%]	1.0 [a]	0.1	0.6 [b]	0.1
pH $_{45}$ - value	6.1	0.09	6.1	0.09
Drip loss [%]	2.8	0.3	3.1	0.3
Reflectance L*	46.4	0.7	46.0	0.7

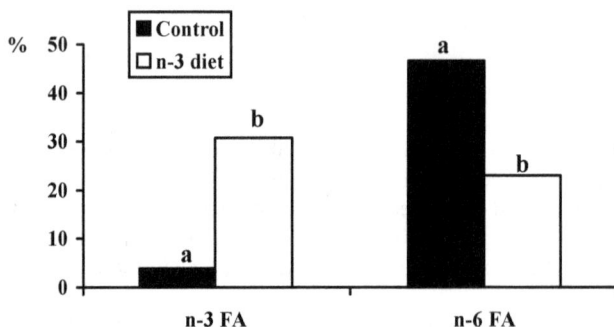

Figure 1   Influence of feeding a diet supplemented with fish oil on fatty acid (FA) composition (wt-%) of phospholipids in total muscle.

The high increase in n-3 FA in skeletal muscle sarcoplasmic reticulum (SR) did not influence the Ca^{2+}-uptake (Nürnberg *et al.*, 1998), but increased the activity of the Ca^{2+}-ATPase signi-ficantly (Figure 2). The discrepancy between unchanged uptake rate and increased ATPase activity (Ca^{2+} - pump activity) with skeletal muscle SR was unexpected. A higher Ca^{2+} -pump activity should result in an increased uptake rate. One reason could be that the vesicles of the n-3 group could be more permeable for Ca^{2+}, but our results show that this is not the case if we take the ratio of ATPase activities with and without a Ca^{2+} - ionophore as a measure for the permeability. It is suggested, that n-3 fatty acid enriched diet can change the complex membrane composition dependent on experimental conditions and animal species leading to different effects on membrane protein activities.

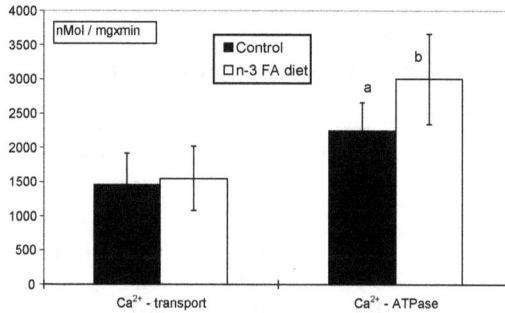

Figure 2 Ca^{2+} - transport and ATPase activity of sarcoplasmic reticulum in longissimus muscle.

Experiment 2: The German Saddle Back pig is known as a pig rich in fatty tissue and with good meat quality, excellent reproduction performance and high vitality. The aim of the study was to quantify the differences in meat and fat quality and the lipid metabolism between German Saddle Back and German Landrace pigs. Saddle Back pigs produced a significant higher proportion of backfat and intramuscular fat and a lower lean meat content compared to GL (Table 2). The meat quality parameters did not differ between homozygous negative (NN) pigs of the two breeds. Heterozygous SB animals indicated higher values for drip loss and reflectance than homozygous SB. More results of carcass quality of these breeds are published by Kuhn et. al (1997, 1998) and Falkenberg et al. (1999).

Table 2   Carcass composition and meat quality.

		German Landrace		German Saddle Back			
		NN		NN		NP	
		LSM	SEM	LSM	SEM	LSM	SEM
Lean meat	%	52.9[a]	0.7	39.8[b]	0.9	42.9[b]	1.3
Fatty tissues	%	28.2[a]	0.8	41.7[b]	1.1	39.1[b]	1.6
Longissimus muscle							
Intramuscular fat	%	1.3[a]	0.1	2.8[b]	0.2	2.9[b]	0.3
pH_{45} - value		6.0	0.1	6.0	0.1	5.9	0.1
Reflectance	L*	47.4	0.6	47.7	0.6	50.1	0.6
Drip loss	%	4.8[a]	0.3	4.1[a]	0.4	6.3[b]	0.6

In both breeds a relationship between intramuscular fat content and fatty acid composition was estimated. The relative concentration of saturated fatty acids was significantly increased in SB pigs (Table 3). The percentage of polyunsaturated fatty acids in SB muscle fat (8.4 %)

was significantly lower than in DL muscle (13.9 %). Palmitic and stearic acids were significantly higher in Saddle Back pigs (Nürnberg *et al.*, 1997). The *de novo* fatty acid synthesis seems to be increased in fatty pigs. The incorporation of saturated fatty acids produced a lower percentage of polyunsaturated fatty acids like linoleic and linolenic fatty acids. The oxidative stability seems to be higher in SB pigs. There was no influence of stress resistance of Saddle Back pigs on fatty acid composition in muscle.

Table 3   Fatty acid composition of longissimus muscle (wt-%).

| | | German Landrace NN | | German Saddle Back | | | |
| | | NN | | NN | | NP | |
		LSM	SEM	LSM	SEM	LSM	SEM
C18:0	%	12.6[a]	0.3	14.7[b]	0.3	14.5[b]	0.4
Saturated fatty acids	%	38.1[a]	0.5	41.3[b]	0.4	40.5[b]	0.7
Unsaturated fatty acids	%	61.9[a]	0.5	58.7[b]	0.4	59.5[b]	0.7
Polyunsaturated FA	%	13.9[a]	0.7	8.4[b]	0.5	8.5[b]	0.9
N-3 FA	%	0.73[a]	0.04	0.36[b]	0.03	0.32[b]	0.05
N-6 FA	%	13.1[a]	0.6	8.0[b]	0.5	8.2[b]	0.8

Experiment 3: Pigs with mutated calcium release channels are stress-susceptible and are known for increased *post mortem* metabolism which results in inferior meat quality. We investigated the meat quality, membrane lipid composition, and lipid peroxidation as well as the calcium transport of the sarcoplasmic reticulum of *longissimus* muscle. Muscle tissue samples were obtained from malignant hyperthermia resistant and susceptible pigs in the time course *post mortem* (0 h, 0.75 h, 4 h, 22 h after death). The incidence of pale, soft, and exsudative meat is higher in MHS pigs, indicating dramatic changes in the *post mortem* metabolism. The lower meat quality scores are reflected by a higher drip loss and a lower pH - value compared to MHR pigs (Table 4).

Table 4   Meat quality of MHS- and MHR-Pietrain pigs.

| | MHS | | MHR | |
	LSM	SEM	LSM	SEM
pH$_{45}$ - value	5.56[a]	0.20	6.27[b]	0.21
Reflectance L*	50.7[a]	3.7	47.4[b]	2.9
Drip loss (%)	5.6[a]	2.2	2.9[b]	1.8

The concentration of total PUFA in muscle phospholipids were not different immediately after death between the two genotypes but declined more rapidly in the MHS muscle *post mortem* (Küchenmeister *et al.*, 1999a). The concentration of long-chain n-3 fatty acids (C20:5, C22:5, C22:6) in the PL fraction of muscle fat is influenced by stress susceptibility (Figure 3). In our study, the concentration of long-chain n-3 fatty acids was significantly higher in MHS pigs. Following these results, it was hypothesised that the PUFAs in MHS muscle undergo an enhanced peroxidation in the *post mortem* time course. Using the malondialdehyde (MDA) concentration, a significantly higher lipid peroxidation in muscle tissues from MHS animals at 45 minutes *post mortem* could be demonstrated (Figure 4). Muscle tissue from MHS pigs oxidised at a significantly higher rate in comparison to tissue from MHR animals (Figure 4, 5).

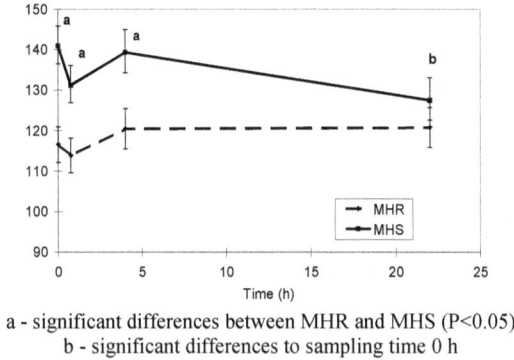

a - significant differences between MHR and MHS (P<0.05)
b - significant differences to sampling time 0 h

Figure 3   N-3 fatty acids (µg/g wt muscle) of phospholipids in muscle.

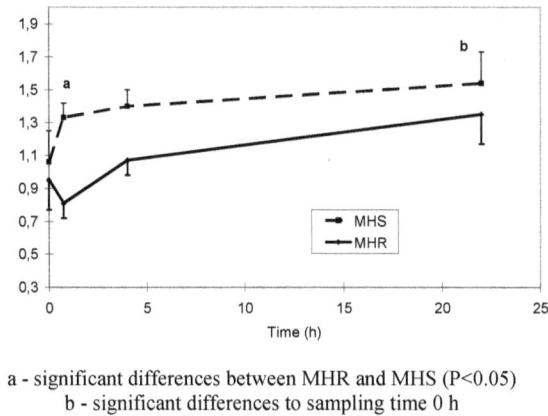

a - significant differences between MHR and MHS (P<0.05)
b - significant differences to sampling time 0 h

Figure 4   Lipidperoxidation of muscle homogenates (malondialdehyde concentration µg/g muscle).

Figure 5 shows decreased resistance to oxidation in muscle homogenates of MHS pigs. The stimulation of peroxidation with Fe^{2+}/ascorbate indicated only small differences in susceptibility to peroxidation in 0 h muscle homogenates of MHS and MHR. However, MHS muscle homogenates obtained at 45 min were more susceptible to peroxidation compared to that of MHR. The increasing susceptibility in the *post mortem* time course can be speculated to be a result of changes in the antioxidative capacity, as also indicated by a decreased vitamin E content (data not shown). An enhanced oxidation of lipids during storage of muscle tissue from MHS animals is suggested. It is concluded that the dramatic changes in meat quality from MHS pigs are possibly caused due to an enhanced oxidation rate additionally to other well known effects. Since this is of crucial importance for meat quality and of potential importance for human nutrition, further attempts to investigate the balance of antioxidant composition and interaction in muscle tissue should be performed.

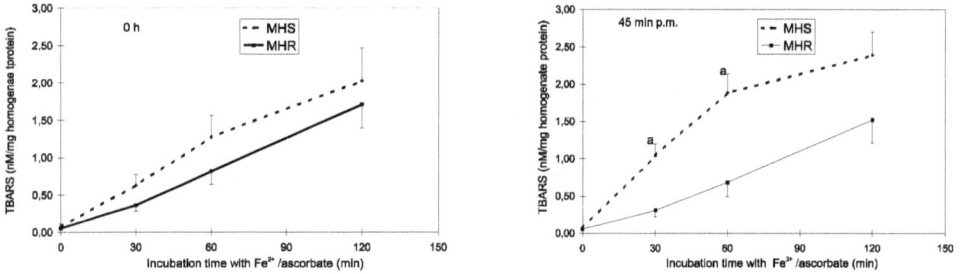

Figure 5   Stimulation of peroxidation by incubation with Fe^{2+} /ascorbate
a-significant differences between MHS and MHR (P<0.05)

The disturbed Ca^{2+} - regulation of the sarcoplasmic reticulum is shown in Figure 6. The ability of SR to sequester Ca^{2+} declined to about 60% in the first 45 min p.m. in MHS homogenates irrespective of CRC state, whereas in MHR samples this decline was about 5%. The CRC can be closed and opened in all samples up to 22 h p.m. and seems to be fully functional at all sampling times. However, the CRC of MHS muscle is almost fully open (CRC not biochemically manipulated), whereas the CRC of MHR pigs is only partially open (Figure 6). The SR-ATPase activity showed a similar decrease (data not shown) suggesting that an inferior ability to take up Ca^{2+} leads to an increased intracellular Ca^{2+} concentration in MHS muscle, activating the *post mortem* metabolism.

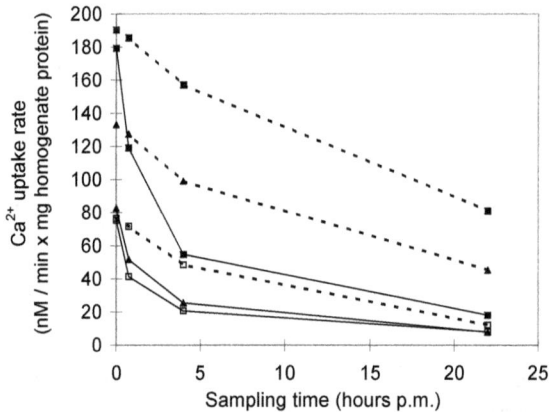

Figure 6   Rates of Ca^{2+} uptake of longissimus homogenates of MHS (——) and MHR (– –)
pigs and different states of the calcium release channel (CRC)
(■), closed CRC; (▲), basic CRC; (□), opened CRC

Conclusions

1. Nutrition with n-3 fatty acid enriched diet effects composition and function of pig muscle membranes but is without effects on meat quality.
2. Pigs with a high content of fatty tissue are no guarantee for good meat quality.
3. Stress susceptible pigs should be excluded from pig breeding programs.

References

Falkenberg, H., G. Kuhn, M. Hartung, M. Langhammer & C. Wolf 1999. Verlauf von biochemischen Kennwerten im Blut von Schweinen mit unterschiedlicher Fettansatzleistung. Arch. Tierzucht, Dummerstorf, 42 (2): 149-159

Küchenmeister, U., K. Nürnberg, I. Fiedler, G. Kuhn, G. Nürnberg & K. Ender, 1999a. Cell injury and meat quality of pig in the time period post mortem from two genotypes susceptible or resistant to malignant hyperthermia. Eur. Food Res. Technol. 209 (2): 97-103

Küchenmeister, U., G. Kuhn, J. Wegner, G. Nürnberg & K. Ender, 1999b. Post mortem changes in Ca2+ transporting proteins of sarcoplasmic reticulum in dependence on malignant hyperthermia status in pigs. Mol. Cell Biochem. 195: 37-46

Kuhn, G., M. Hartung, H. Falkenberg, G. Nürnberg, M. Schwerin & K. Ender, 1997. Wachstum, Körperzusammensetzung und Fleischbeschaffenheit von im Fettansatz genetisch differenten Schweinen, Arch. Tierzucht, Dummerstorf, 40 (4): 345-55

Kuhn, G., M. Hartung, K. Nürnberg, I. Fiedler, H. Falkenberg, G. Nürnberg & K. Ender, 1998. Körperzusammensetzung und Muskelstruktur von genetisch differenten Schweinen, Arch. Tierzucht, Dummerstorf, 41 (6): 589-596

Nürnberg, K.; G. Kuhn, K. Ender, G. Nürnberg & M. Hartung, 1997. Characteristics of carcass composition, fat metabolism and meat quality of genetically different pigs. Fett/Lipid 99 (12): 443-446

Nürnberg, K.; U. Küchenmeister, K. Ender, G. Nürnberg & W. Hackl, 1998. Influence of dietary n-3 fatty acids on the membrane properties of skeletal muscle in pigs. Fett / Lipid 100 (8): 353-358

Otten, W., 1993. Untersuchungen von Zellmembranlipiden und Metaboliten des Kohlenhydrat- und Lipidstoffwechsels bei Schweinen mit genetisch unterschiedlicher Stressresistenz. Diss., TU München

Food waste products in diets for growing-finishing pigs

N. P. Kjos & M. Øverland

Department of Animal Science, Agricultural University of Norway,
P. O. Box 5025, N-1432 Ås, Norway

Summary

Forty-eight growing-finishing pigs were used to study the effect of food waste products in diets. Increasing dietary levels reduced average daily intake of feed dry matter and feed dry matter/gain ratio, reduced fat firmness, and reduced lightness (L*) of backfat and loin meat. The proportion of saturated and polyunsaturated fatty acids in backfat decreased and increased, respectively. Sensory quality of loin muscle was not affected. Food waste products could be used in diets for growing-finishing pigs when used in moderate levels.

Keywords: Food waste products, Pigs, Growth performance, Carcass merits, Meat quality

Introduction

Food waste products can be defined as the edible wastes from food production and consumption, and may constitute as much as 20% of the total human food supply (Westendorf *et al.*, 1998). The increased cost for disposal of such products in landfills, and governmental restrictions, have led to growing interest in using food waste in diets for pigs. Successful use of these products as feed is also of benefit for the pig farmers, because of low cost.

In order to even out the great variation found in chemical composition and nutritive value within many of the organic waste products, several of these products are mixed together in a liquid food waste feed with a defined nutritive quality (FW). To ensure good hygienic quality, ingredients such as food leftovers, food waste products containing meat, blood and some dairy waste products should be sterilised for at least 20 minutes at 133°C (Royal Norwegian Ministry of Agriculture, 1979). The remaining ingredients are added after the sterilisation process, and the liquid feed mixture is heated to 70°C before it is delivered to the pig farmer.

The present experiment was carried out to determine the optimum dietary level of FW in a wet feed with respect to 1) growth performance and carcass quality of growing-finishing pigs and 2) meat quality, including sensory quality.

Material and methods

The experiment was carried out with 48 crossbred pigs from eight litters, allotted to six dietary treatments on the basis of initial weight, sex and litter. Average initial weight was 28.3 kg, SD 4.8 kg. The pigs were kept in six partially slatted pens designed for individual feeding, and they were individually fed twice daily according to a restricted feeding scale. The diets were mixed with water to obtain a liquid feed of 20% dry matter (DM). Two wet feed mixes (FW1 and FW2), and five barley-soybean meal based diets (SB00, SB20, SB40, SB60 and SB80), were used. The FW consisted of (as feed basis): food leftover products (324 g kg^{-1}); pizza factory waste (47 g kg^{-1}); blood from slaughterhouse (86 g kg^{-1}); ice cream factory waste (20 g kg^{-1}); dairy waste (92 g kg^{-1}); bakery waste (122 g kg^{-1}); potato waste (278 g kg^{-1}); others (31 g kg^{-1}). The experimental diets were:

Diet FW0   - Control diet with 100% of FUp as diet SB00.
Diet FW20   - Diet with 20% of FUp as FW1, and 80% as SB20.
Diet FW40   - Diet with 40% of FUp as FW1, and 60% as SB40.
Diet FW60   - Diet with 60% of FUp as FW1, and 40% as SB60.
Diet FW80   - Diet with 80% of FUp as FW1, and 20% as SB80.
Diet FW100 - Diet with 100% of FUp as FW2.

The inclusion of FW in diets FW20, FW40, FW 60 and FW80 was 438, 678, 824, and 924 g kg^{-1} (as fed), respectively.

Four pigs from each treatment were slaughtered at day 84 or at day 91 of the experiment. Carcass characteristics were measured one day after slaughter. Each of the loin muscles were dissected for chemical and sensory analyses, and samples of backfat were taken for analyses of fatty acid composition. Sensory analysis was conducted on fresh loin muscle (fresh) and on loin muscle stored at -16°C for six months (stored). The sensory analysis was conducted at the Norwegian Food Research Institute, Ås, according to international standards (ISO 3972 Sensory analysis - Methodology - Method of investigating sensitivity of taste).

Growth performance, carcass characteristics, sensory quality and fatty acid composition in backfat were analysed using PROC GLM in SAS. In the tables, levels of significance is indicated as NS = not significant, P<0.05, P<0.01, and P < 0.001.

Results and discussion

The FW were characterised by a low DM content of 21.4%, a relatively high fat content of 3.3%, a low crude fibre content of 0.5%, and a high content of sodium (0.1%) and chloride (0.2%). Further, FW contained 4.9% crude protein, 11.5% N-free extracts, 1.2% ash, and 0.33 kg^{-1} FUp. The pH was 3.9. The proportions of saturated (SAFA), monounsaturated (MUFA) and polyunsaturated (PUFA) fatty acids in relation to total content of fatty acids were 42%, 38% and 18%, respectively. Increasing levels of FW in the diet resulted in increased contents of crude fat, crude protein, ash, calcium, sodium and chloride, while the contents of N-free extracts and crude fiber decreased. The proportions of the fatty acids C14:0, C16:0, C16:1, C18:1, C20:5 (EPA) and C22:6 (DHA) increased, while that of C18:0 and C18:2 decreased. In total, the proportions of SAFA and MUFA increased, while that of PUFA decreased, with increasing levels of FW. The results are described more closely by Kjos et al., (2000).

Table 1. Growth performance (number of pigs per treatment = 8).

Diet	FW00	FW20	FW40	FW60	FW80	FW100	P <	SEM[1]
Final weight, kg	107.8	107.5	103.8	103.1	101.8	106.7	NS	2.64
Average daily gain, g	908	906	860	862	841	892	NS	25.4
Feed unit pig day^{-1}	2.30[a]	2.30[a]	2.20[ab]	2.12[ab]	2.08[b]	2.11[ab]	0.001	0.047
Feed/gain, kg DM kg^{-1}	2.03[a]	2.03[a]	2.01[a]	1.91[a]	1.90[a]	1.74[b]	0.001	0.041
Feed unit pig/kg gain	2.53	2.53	2.56	2.47	2.50	2.36	NS	0.053

[1] Standard error of the mean.
[a, b] - Different letters indicate significant difference among dietary treatments (P < 0.05).

Growth performance is shown in Table 1. Increasing dietary levels of FW reduced average daily feed DM intake (ADFI), average daily intake of FUp and feed DM/gain ratio (linear, P < 0.001). Pigs fed the diets with 60% FW or more had a lower ADFI compared with pigs fed the rest of the diets. When expressed as FUp, only the diet FW80 had a significantly lower intake. Feed DM/gain was not affected by the FW level, except for diet FW100. Thus, FW had no adverse effect on growth performance of growing-finishing pigs when used in moderate levels (20 - 50% of the total FUp content).

Carcass merits, meat quality, and some sensory attributes, are given in Table 2. Lean percentage and meat score in carcass were not significantly affected by dietary level of FW. There was an effect of FW level on P2 backfat thickness (P < 0.03), with the highest and lowest values observed for FW80 and FW100. Fat firmness was linearly reduced (P < 0.02) when dietary FW increased. The adverse effect on fat firmness of dietary FW levels higher than 60% of FUp was probably an effect of increased dietary fat level and daily intake of PUFA. Dietary FW level affected drip loss of meat (P < 0.05), with the highest value for diet FW80. Lightness of subcutaneous fat and meat (L* values) decreased linearly (P < 0.01) with increasing levels of FW. This indicated that fat and loin meat from pigs fed diets FW80 and FW100 were darker than the pigs fed the control diet. The differences were probably not large enough to adversely affect meat quality. Redness (a*) and yellowness (b*) of fat and meat were not affected. Adding of FW to diets had no effect on thiobarbituric acid value (TBA), or on sensory attributes, of fresh or stored meat. Numerically, six months storage gave higher intensities for rancid odour, rancid taste, off odour and off taste compared to fresh meat.

Table 2.   Carcass merits and meat quality.

Diet	FW00	FW20	FW40	FW60	FW80	FW100	P <	SEM
P2 backfat thickness, mm	11.9[ab]	11.6[ab]	12.9[ab]	12.4[ab]	14.9[a]	10.7[b]	0.05	0.87
Fat firmness (1 - 15)	11.8[a]	11.7[ab]	11.2[ab]	11.0[ab]	10.9[ab]	10.8[b]	0.001	0.21
Lean, %	53.1	52.8	52.6	52.5	52.7	54.0	NS	0.63
Carcass weight, kg	77.4	78.0	74.7	74.0	74.9	75.1	NS	2.41
Dressing %	71.8	72.5	71.8	71.7	73.2	70.3	NS	0.80
Drip loss of meat, %	3.2[a]	4.0[ab]	3.8[ab]	3.5[ab]	4.2[b]	3.6[ab]	0.05	0.20
Fat lightness, L*	79.1[a]	79.3[a]	78.8[a]	78.0[ab]	76.6[b]	75.8[b]	0.001	0.68
Fat redness, a*	3.7	3.5	4.5	4.1	4.3	4.3	NS	0.51
Fat yellowness, b*	4.1	4.4	4.5	4.5	4.7	4.5	NS	0.35
Meat lightness, L*	60.6[a]	59.2[ab]	60.0[ab]	58.9[ab]	58.1[ab]	57.8[b]	0.01	0.74
Meat redness, a*	6.8	7.9	8.3	7.9	7.6	7.6	NS	0.55
Meat yellowness, b*	3.6	3.8	4.1	3.7	3.5	3.5	NS	0.26
TBA, fresh sample	189.0	186.0	182.9	196.2	185.0	222.8	NS	21.10
TBA, stored six months	240.9	234.4	275.6	281.8	266.1	287.0	NS	21.60
Sensory analysis, loin muscle[1]								
Fresh, off taste (1 - 9)	3.26	2.77	3.20	2.87	2.89	3.16	NS	0.179
Fresh, rancid taste (1 - 9)	1.25	1.14	1.16	1.11	1.21	1.25	NS	0.048
Stored, off taste (1 - 9)	3.42	3.13	3.67	3.37	3.34	3.54	NS	0.170
Stored, rancid taste (1 - 9)	2.45	2.36	2.81	2.66	2.68	2.82	NS	0.210

[1] Sensory attributes judged on a scale from 1 to 9, where 1 is no intensity and 9 is high intensity of the attribute.
a,b - different letters indicate significant differences among experimental diets (P<0.05).

Table 3. Fatty acid composition (g/100 g fatty acid) in subcutaneous fat.

Diet	FW00	FW20	FW40	FW60	FW80	FW100	P <	SEM
C22:5+C22:6, n-3	0.08[a]	0.05[a]	0.37[b]	0.29[ab]	0.50[b]	0.54[b]	0.001	0.08
SAFA	42.00[a]	41.18[ab]	39.32[bc]	37.66[c]	37.30[c]	36.71[c]	0.001	0.84
MUFA	46.00	46.29	46.22	46.27	46.89	47.29	NS	0.55
PUFA	11.98[a]	12.54[a]	14.45[b]	16.06[b]	15.81[b]	15.99[b]	0.001	0.08

a,b,c - different letters indicate significant differences among experimental diets (P<0.05).

Table 3 shows that dietary FW exerted a marked effect on the fatty acid composition of backfat. The proportion of SAFA decreased (linear, $P < 0.001$) and the proportion of PUFA increased (quadratic, $P < 0.001$) with increasing FW levels. The proportions of PUFA were significantly higher ($P < 0.05$) for the diets with 40% or more FW on FUp-basis. Several other studies, amongst others Øverland *et al.,* (1996) and Kjos *et al.,* (1999), have also shown that dietary fatty acid composition influenced the fatty acid composition in pork.

Diets causing increased proportions of unsaturated fatty acids (UFA) in pork lipids could be expected to increase susceptibility to lipid oxidation. Although proportions of UFA in backfat of the pigs fed FW were significantly higher than those fed the control diet, no appreciable differences in TBA or sensory attributes were found among dietary treatments.

Conclusions

To conclude, FW could be used in diets for growing-finishing pigs without adversely affecting growth performance, carcass merits and sensory quality when used in moderate levels. Optimal dietary inclusion level of FW was in the range of 20 to 50% of FUp.

References

Kjos, N. P., A. Skrede & M. Øverland, 1999. Effect of dietary fish silage and fish fat on growth performance and sensory quality of growing-finishing pigs. Can. J. Anim. Sci. 79: 139-147

Kjos, N. P., M. Øverland, E. Arnkværn Bryhni & O. Sørheim, 2000. Food waste products in diets for growing-finishing pigs. Effect on growth performance, carcass characteristics and meat quality. Submitted to Acta Agric. Scand. Sect. A, Animal Sci (in press)

Øverland, M., O. Taugbøl, A. Haug & F. Sundstøl., 1996. Effect of fish oil on growth performance, carcass characteristics, sensory parameters, and fatty acid composition in pigs. Acta Agric. Scand. Sect. A, Animal Sci. 46: 11-17

Westendorf, M. L., Z. C. Dong & P. A. Schoknect, 1998. Recycled cafeteria food waste as a feed for swine: Nutrient content, digestibility, growth and meat quality. J. Anim: Sci. 76: 2976-2983

Conservation and development of the Bísaro pig. Characterisation and zootechnical evaluation of the breed for production and genetic management[⊗]

J. Santos e Silva[*][1], *J. Ferreira-Cardoso*[2], *A. Bernardo*[3] *& J. S. Pires da Costa*[4].

[1]*DREDM - EEPA, Quinta do Pinhó, S. Torcato, 4800 Guimarães.*
[2]*UTAD - SEBA, apt. 202, 5000 Vila Real.*
[3]*DRATM - CEB, Quinta da Veiga, 5470, Montalegre.*
[4]*EZN (INIA) - Estação Zootécnica Nacional, Fonte Boa, Vale de Santarém, 2000.Portugal.*

Summary

This study is part of an extensive plan of conservation and recovery of an ancestral genetic resource, the Bísaro pig, and has the objective of reversing the lack of variability that has been observed for the past 50 years, as well as establishing technical and economic references which are necessary for the development of high quality production involving this breed and its production systems. This study shows that the Bísaro breed at 105 kilograms liveweight maintains the zootechnical characteristics of its ancestors: slow growth (550 g day^{-1}), low feed efficiency (3.77), poor conformation (U), low prime cut yield, high bone yield, average quantity of external fat (20 mm) and a good percentage of 2.64 for intramuscular fat best *post mortem* pH values and water retention capacity than other breeds studied at the same time. The criteria for breed conservation and breeder selection are: molecular identification of the halothane gene, CRC locus, (revealing the secretion of allele n (Nn) which has negative effects on sensorial quality and meat technology), mean values of *post mortem* pH observed in the families, as well as ADG, FCR and backfat.

Keywords: Conservation, Recovery, Breed, Variability & Quality.

Introduction

All over Europe pig production has tended towards standardised production systems with similar facilities used, feeding methods and animal genetics and a reduction in available area. At the same time, in Portugal a decline was observed in the Alentejano and Bísaro national breeds (the last one almost extinct) which are less productive and poorly adapted to intensive farming. The unpredictable evolution in a medium/long term of pig production, considering the unknown evolution of the consumption in what concerns quality and methods of production, together with the aim of improving performance and productivity, imply the existence of available genetic resources and their safeguard (Hammond & Leitch, 1998). The Bísaro breed can be one of these resources. This genetic resource has been spread all over Portugal mainly from the river Tagus to the north of Galiza in Spain, but became almost extinct in the last decade (Santos e Silva *et al.*, 1996). This breed had a poor growth rate and bad conformation, excessive adult weight and body size, but it has traditionally been known for its excellent meat quality, suitable for processing (Lima, 1865; Pinto 1878; Janeiro, 1944). The current study is integrated in a larger programme[⊗] to recover the Bísaro pig and intends to be a preliminary productive characterisation of this breed, of which the first phase of the

[⊗] PAMAF (7173) INIA – Programa de Apoio à Modernização Agrícola e Florestal (Ministry of Agriculture). Protection Recover and Development of " Bísaro " pig. Coordination Prof. José Pires da Costa.

inventory and identification has been realised (Santos e Silva *et al.*, 1998). Growth and carcass performances were evaluated in Bísaro pigs in comparison with the other Portuguese breed (Alentejana) and a Landrace X Large White cross common in Portugal, in an attempt to define criteria for breed conservation and use in improvement programmes.

Material and methods

Origin of animals and genetic types

In this experiment 48 pigs (12 entire male pigs, 12 castrated males and 24 females) of nearly 30 kilograms liveweight were used. Piglets were produced by mating 8 Bísaro sows with 5 boars genealogically identified in the three conservation nucleii of Montalegre, São Torcato and UTAD considered to be representative samples of the genetic types of the breed found in the Trás-os-Montes and Entre Douro e Minho regions. The same animals were evaluated for carcass and meat quality at 100 kilograms liveweight in comparison with 6 F_1 Landrace*Large White (LR*LW) and 6 Alentejano (Al) pigs kept at the same time and in the same system of housing and feeding.

Growth performance

Performance tests were carried out on pigs of between 30 and 100 kilograms at EZN research centre (Estação Zootécnica Nacional [24 Bísaros]) and at UTAD (University of Trás-os-Montes [24 Bísaros]). All pigs received the same diet containing 12 MJ EM/kg and approximately 16% CP. The method of feeding used was the semi-*ad libitum* method in practice at EZN (Pires da Costa, 1974). Performance and carcass traits were evaluated at two different phases, which coincided with two different body weights: one (phase 1) at 105 kg and the other (phase 2) at 150 kg.

Slaughter Procedure and Processing

When the liveweight of the pigs on performance test (EZN: 6 Bísaros, 6 LR*LW, 6 Al) and UTAD (6 Bísaros) was approximately 100 kilograms they were sent to the abattoir and slaughtered after 24 hours of resting. All carcasses were weighed and stored for 24 hours in a cold room; pH measurements were taken in the *Longissimus Dorsi* (LD) and *Semi-membranosus* (SM) between 45 and 60 minutes *post mortem* and 24 hours after killing. On the following day, linear fat and loin area determination were taken on the left half side of the carcass. The resulting joints were weighed (shoulder, ham, loin with bone, belly and head) and measurements of the carcass and ham were taken. Fat and muscle samples were collected for determination of intramuscular fat content (ISO, 1973), colour (reflectance CIE L*), water retention capacity, drip and cooking loss (Honikel, 1987) and chemical analyses, dry matter, ash, protein (AOAC, 1990). Heavy Bísaro pigs ± 150 kg (Phase 2) were sent to the same slaughter house and an identical procedure of slaughter and processing was used.

Genetic test for halothane sensitivity

All pigs in the experiment were tested for the halothane gene with genotypes determined by analysing the mutation in porcine skeletal muscle calcium release channel (CRC) gene, as described by Brenig & Brem (1992).

Statistical analysis

Data of performance test, carcass and meat quality were subjected to analysis of variance applying the general linear model (SAS programme) to evaluate effects of lines, sex and breed. Data of pH_{45m} and pH_{24h} were subject to a similar analyses including the effect of genotype of CRC gene, slaughter weight, distance to the abattoir and interaction genotype (Hal) x slaughter weight. When the effect of the sources of variation was significant ($P < 0.05$), means were compared by the LSD test.

Results and discussion

Performance data

Table 1.   On-farm (EZN) - Performance test by phase (1 and 2). Montalegre / S. Torcato Lines.

	Conservation nucleii		Mean
	Montalegre	S. Torcato	
N	12	12	24
Initial weight kg	32.8 ± 6.1 [a]	26.2 ± 6.2 [b]	29.8 ± 7.2
Final Weight$_1$ kg	100.9 ± 14.1 [a]	89.7 ± 12.3 [a]	95.8 ± 14.8
Final weight$_2$ kg	149.8 ± 21.1 [a]	138.7 ± 13.0 [a]	144.3 ± 19.1
ADG$_1$ kg day^{-1}	0.577 ± 0.08 [a]	0.538 ± 0.115 [a]	0.559 ± 0.102
ADG$_2$ kg day^{-1}	0.544 ± 0.109 [a]	0.514 ± 0.153 [a]	0.534 ± 0.140
Average daily food intake$_1$ g day^{-1}	2166 ± 163 [a]	1971 ± 221 [a]	2052 ± 23
Average daily food intake$_2$ g day^{-1}	2617 ± 133 [a]	2620 ± 128 [a]	2618 ± 13
FCR$_1$ kg/kg	3.81 ± 0.53 [a]	3.71 ± 0.86 [a]	3.77 ± 0.72
FCR$_2$ kg/kg	4.92 ± 0.98 [a]	5.99 ± 3.08 [a]	5.45 ± 2.44

Means with the same letter are not significantly different (P > 0.05).
ADG – Average daily gain in phases 1 and 2; FCR – Food conversion ratio in phases 1 and 2.

Means for On-Farm Performance are shown in table 1. There were no significant differences between the two lines for all the traits analysed, and the results indicate that Bísaro pigs have a slow growth rate and which does not differ significantly between the interval from 30 to 96 kg (0.559 kg day^{-1}) and from 96 to 150 kilograms liveweight (0.534 kg day^{-1}). However, growth during the first phase proved to have a better feed efficiency (feed conversion ratio of 3.77) than the second phase of the trial (FCR=5.45). These values show the low growth rate and feed efficiency of this breed in relation to LR*LW.

Carcass compositions, linear measurement and dissection

Table 2.   Carcass measurement. Differences between breeds.

	Breeds		
Trait	Bísaro	LR x LW	Alent.
N	12	6	6
Weight at slaughter kg	105.5 ± 5.9 [a]	106.8 ± 3.5 [a]	96.0 ± 4.4 [b]
Cold Carcass weight kg	81.2 ± 4.6 [ab]	84.3 ± 5.0 [a]	75.8 ± 4.4 [b]
Carcass yield %	77.0 ± 1.8	78.8 ± 3.0	78.9 ± 1.1
Carcass drying %	2.3 ± 0.2	3.5 ± 0.4	1.9 ± 0.6
Conformation	U	E	R
% Ham	27.0 ± 1.0 [b]	29.2 ± 1.1 [c]	25.4 ± 1.1 [a]
% Shoulder	20.5 ± 1.0 [b]	19.8 ± 0.6 [ab]	18.9 ± 0.9 [a]
% loin	24.4 ± 1.1	25.6 ± 1.5	25.0 ± 2.2
% Belly	19.1 ± 0.9 [b]	17.5 ± 0.7 [a]	22.5 ± 1.1 [c]
% Head	8.5 ± 0.7 [b]	7.4 ± 0.3 [a]	7.5 ± 0.9 [a]
Carcass length cm	106.9 ± 4.2 [c]	97.6 ± 4.1 [b]	90.2 ± 1.0 [a]
Carcass width (breast)	37.6 ± 1.5	37.1 ± 1.0	38.7 ± 3.2
Carcass width (waist)	26.2 ± 2.4	28.3 ± 1.6	27.0 ± 3.0
Ham length cm	84.2 ± 3.1 [b]	76.9 ± 1.6 [a]	78.5 ± 1.5 [a]
Ham width cm	31.8 ± 1.1 [b]	30.4 ± 1.6 [a]	30.7 ± 1.0 [ab]
Loin length cm	91.6 ± 4.6 [c]	85.3 ± 3.0 [b]	73.5 ± 0.5 [a]
Backfat thickness cm			
P_1	18.8 ± 7.0 [a]	14.9 ± 2.2 [a]	53.7 ± 4.5 [b]
P_2	19.5 ± 7.4 [a]	14.3 ± 1.7 [a]	54.4 ± 3.8 [b]
P_3	21.1 ± 4.5 [a]	15.6 ± 1.1 [a]	52.2 ± 2.6 [b]
Loin muscle area (cm^2)	29.3 ± 4.5 [b]	45.4 ± 5.0 [c]	18.9 ± 0.9 [a]

A summary of the results is given in table 2. Carcass yield did not differ significantly (P > 0.05) between breeds. Linear measurements on carcass, ham and loin show that the Bísaro breed has a longer carcass, loin and ham (P< 0.05) when compared with the LR*LW line and Alentejano breed. Generally, with the joints obtained after dissection, the Bísaro presents a lower yield in first category cuts when compared with LR*LW but is better than the Alentejano. Conversely, the Bísaro breed has a significantly higher yield of third category cuts (head 8.5 % and belly 19 %) in relation to de LW*LR (7.4 % and 17.5 % respectively). Values for subcutaneous fat measured at the 13th and 14th vertebra at P_1, P_2 and P_3 points were significantly different (P<0.05) between breeds. Although it did not present excessive fat, the Bísaro breed had a 19.8 mm average of backfat, which places it between the LR*LW pigs (14.9 mm) and the Alentejana breed (53.6 mm). The loin muscle area was larger in the Bísaro (29 cm^2) than in Alentejana (19 cm^2) (P < 0.05) but smaller than in F$_1$ LR*LW pigs (45 cm^2). Applying the actual EU system for carcass classification in Portugal the three breeds will be ranked in the following position: LR*LW (E), Bísaro (U) and Alentejano (R), however these rules can' t be applied to Portuguese breeds as they are only justified for high quality meat production. The results obtained in this experiment are in agreement with the result reported by Lima (1886) and Janeiro (1944).

Meat quality measurement

Table 3. Meat quality measurement. Differences between breeds.

Trait	Breeds		
	Bísaro	LR x LW	Alent.
N	12	6	6
PH$_{45}$ SM	5.72 ± 0.30 [a]	5.57 ± 0.29 [a]	5.44 ± 0.11 [a]
PH$_{45}$ LD	5.95 ± 0.48 [b]	5.35 ± 0.04 [a]	5.89 ± 0.41 [ab]
PH$_{ul}$ SM	5.67 ± 0.28 [b]	5.40 ± 0.07 [a]	5.52 ± 0.02 [ab]
PH$_{ul}$ LD	5.56 ± 0.15 [b]	5.37 ± 0.04 [a]	5.39 ± 0.05 [a]
Reflectance L* (LD)	54.02 ± 3.45 [ab]	56.98 ± 1.88 [b]	50.67 ± 2.14 [a]
Drip loss (LD) %	1.49 ± 0.67 [a]	3.63 ± 0.37 [b]	2.31 ± 0.90 [a]
Cooking loss (%)	14.40 ± 1.73 [a]	20.61 ± 0.36 [b]	16.29 ± 2.08 [a]
Tenderness WBSF (kg/cm^2)	9.26 ± 1.41 [b]	9.19 ± 1.04 [ab]	7.28 ± 1.47 [a]
% Water	73.21 ± 0.90	73.73 ± 0.36	72.67 ± 0.50
% Ash	1.15 ± 0.03	1.17 ± 0.04	1.15 ± 0.03
% Protein	22.98 ± 0.92 [b]	23.87 ± 0.22 [c]	21.82 ± 0.27 [a]
Intramuscular fat content %	2.64 ± 1.39 [b]	0.84 ± 0.25 [a]	4.05 ± 0.41 [b]

In table 3, means for meat quality measurements show that the Bísaro pigs have significantly better values for pH$_{45m}$, pH$_{24h}$, colour (CIE L*), drip and cooking losses than LR*LW pigs (P < 0.05). This suggests a high sensorial and technological quality of fresh meat contributing to the large acceptability of this meat and for processing of high quality pork products. When compared with LR*LW, Bísaro pigs also show higher intramuscular fat (0.84 and 2.64, respectively).

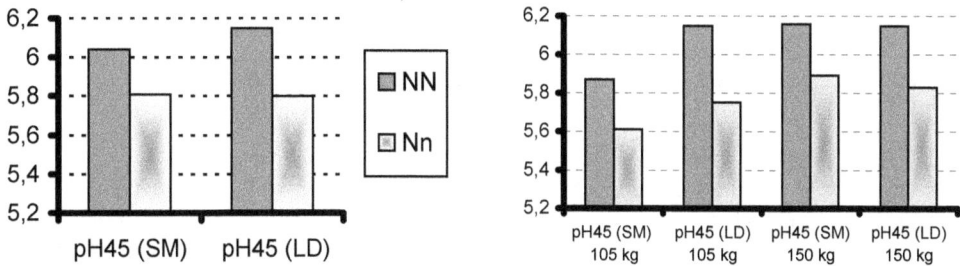

Figure 1. Influence of CRC genotypes on velocity of pH fall.

According to the Halothane sensitivity gene (CRC) this study (fig. 1) confirms what has already been reported by other authors (Monin *et al.,* 1999), namely the faster fall in pH *post mortem* for pigs carrying the allele n (heterozygous Nn) relative to homozygous pigs (NN) (P < 0.05) whatever the finishing weight that is being considered. However no significant differences were observed between conservation nucleii of Bísaro, sexes and testing farms (EZN and UTAD).

Phenotypic of boars offspring

Table 4. Mean phenotypic values per boar.

| | | | | Mean deviations | | Mean |
| | | | | **Offspring** | | |
Boar	N Descendants	ADG kg/day	FCR kg/kg	Backfat Thickness P_1 mm	PH_{45min}, LD	PH_{24h}, LD
604	10	-0.021	-0.06	-2.00	6.12	5.52
607	2	-0.040	0.35	-3.26	5.84	5.44
B157	4	-0.040	-0.01	-4.52	6.27	5.52
B65	2	0.051	-0.16	-1.65	5.36	5.47
PB	4	0.067	-0.22	6.90	5.55	5.42
C.V.		18.2%	19 %	35.8 %	7.1 %	3.4 %

The results in table 4 together with the characterisation of the CRC genotype should be an interesting starting point for future work in the selection index for this breed. The high heritability of backfat (0.52 in semi-*ad libitum* condition, Stewart and Schinckel, 1989) and the large variability (35.8 %) shown for this criteria, suggest that this characteristic should be very vulnerable in selection programmes.

On the other hand the means per boar (ancestors of piglets tested [table 4]) show Boars (B65, PB) with a very low pH_{45m} (5.36 and 5.55), a possible indication of some genetic variability. pH after 24 hours showed low variation with values close to acid meat.

Distribution of Halothane gene in Bísaro pigs in different nucleus

Table 5. Genotype frequencies for CRC gene by nucleus.

| | | Genotype | | | |
	Num.	NN	Nn	Nn	unknown
São Torcato	7	5	0	0	2
Montalegre	8	3	3	0	2
UTAD	15	4	11	0	0
Total	30	12	14	0	4

Using the technology for identification of the halothane gene (table 5) we can see that pigs carrying the allele n (genotypic frequency Nn 54 %) were bred in UTAD and Montalegre while pigs born in S. Torcato are free of this allele. Historically the Bísaro pig was described as a breed having poor performance and bad carcass conformation but had never been described as a breed suitable for the malignant hyperthermia. This result could indicate some introduction of exotic genotypes and may be the most important step in our study to be reached in future research clarifying our preliminary work and saving the purity of this breed. Future studies using a larger sample from unrelated individuals should be carried out in order to estimate the allelic distribution in this population.

Conclusions

The results of this preliminary study suggest that the pigs used in this experiment are similar to the ancestral Bísaro pigs: slow growth rate (550 g/day up to 100 kg liveweight), low feed conversion (3.77), bad conformation, medium quantity of subcutaneous fat (20 mm) and favourable values for traits which influence sensorial quality and meat technology. Variability found in different productive characteristics suggests that the breed can be improved. The molecular identification on tested animals for the halothane gene (locus CRC), which revealed the segregation of allele n (Nn) – negative for sensorial quality and meat technology – and the mean values of *post mortem* pH (decrease of velocity and amplitude) observed in some families, could be some of the criteria for the selection of breeders. However, when we consider the need for preserving intra-breed variability while keeping the high quality of the Bísaro meat, we think that the management and genetic development of this breed should be carried out in two ways: 1. cryogenic conservation of semen with the preservation of productive variability; 2. development of two genetic lines (1-fat line, 2-lean line). From our preliminary results we can conclude that much more work has to be done namely in the field of the genetic identification of different lines of pigs today offered to the farmers as Bísaros.

Acknowledgements
We are grateful to Eng. A. Pereira and Dr. Nuno Ferrand (Centro de Estudos Ciência Animal ICETA-University of Porto) for assistance in the analyses of DNA and J.A. Silva (University of Trás-os-Montes) for the meat chemical analysis. We also thank Dr. Telo da Gama and one anonymous referees for reading a draft of this paper and help on genetic discussion.

References

AOAC, 1990. Association of Official Analytical Chemists. Official methods of analysis. (15th ed.). Washington, D.C.

Brenig B. & G. Brem, 1992. Genomic organization and analysis of the 5' end of porcine ryanodine receptor gene (ryr 1). FEBS letters 298 (2): 277-279

Hammond K. & H.W. Leitch, 1998. Genetic resources and global programme for their Management. The genetics of the pig (M.F. Rothschild and Ruvinsky) 14: 405-425

Honikel, K.O., 1987. How to measure the water-holding capacity of meat? Recommendation of standardized methods. In: Evaluation and control of meat quality in pigs, P.V. Tarrant, G. Eikelenboom & G. Monin (editors), Martinus Nijhol Publ.:, 129-142

ISO, 1973. Determination of free fat content in meat products (ISSO 1444-1973). Geneva, International Organisation for Standardisation

Janeiro, J.P., 1944. A suinicultura em Portugal. In Boletim Pecuário nº2, Ano XII, 194 pp.

Lima, S.B. (1865) História natural e económica do porco, classificação e indicação das espécies e raças suínas, Arquivo rural, 3: 91-99

Monin, G., C. Larzul, P. Le Roy, J. Culioli, S. Rousset-Akrim, A. Talmant, C. Touraille & P. Sellier, 1999. Effects of the Halothane genotype and slaughter weight on texture of pork. J. Anim. Sci., 77: 408-415

Pires da Costa, J., 1974. Effects of dietary protein levels and energy intake on performance, carcass quality and protein utilization by growing pigs. PH.D. Thesis National Institut for research dairying, Reading 221 pp.

Pinto, J.F.M., 1878. Zootechnia dos animales suinos. In Vet. Zoo. Dom; Coimbra Uni: 440 pp

Santos e Silva, J., 1996. O porco bísaro em extinção: contributo para a sua identificação, caracterização e recuperação. Veterinária técnica:3: 20 26

Santos e Silva, J., Bernardo A. & J. Pires da Costa, 1998. Caracterização e inventário genético do porco bísaro, com recurso a genes de efeito visível. Sua utilização na constituição de um núcleo de conservação genética "In Vivo". IV Simpósio Internacional do porco mediterrânico. Èvora. Portugal.

Stewart, TS. & A.P. Schinckel, 1989. Genetics parameters for swine growth and carcass traits. In Young, L.D. (ed.) Genetics of Swine. USDA-EAS, Clay Center, Nebraska: 77-79

Meat consumption and cancer risk

M. Zimmermann

Laboratory for Human Nutrition, Institute of Food Science, ETH Zürich, 8092 Zürich, Switzerland

Introduction

It is estimated that dietary factors play a causative role in about one-third of all cancers (American Cancer Society, 1998). There is a widespread perception among the public that meat intake increases risk of cancer, particularly cancer of the colon and the stomach (American Institute for Cancer Research, 1997). However, the epidemiologic data to support these putative associations are inconsistent, and the significance of meat's role as a cancer risk factor remains a subject of debate (Baghurst *et al.*, 1997; American Institute for Cancer Research, 1997). This review summarises new epidemiologic evidence presented within the past few years on this important question.

It is worth remembering that epidemiologic associations between dietary components and chronic disease do not establish cause and effect relationships (Friedman, 1994). This may be particularly true for diet and cancer. Diet-related variables are numerous, complex and interdependent, and it is often difficult to single out specific dietary factors. Epidemiologic data need to be confirmed by testing in animal and metabolic studies, and, ultimately, clinical trials are necessary to establish causal relationships (Langseth, 1996).

Keywords: Meat, Saturated fat, Cancer, Colon, Stomach

Meat Intake and Cancer

Meat consumption has been linked to increased risk for colon cancer in both women (Willett *et al.*, 1990) and men (Giovanucci *et al.*, 1994). In the prospective study by Willett *et al.* (1990) among 88,751 women 34 to 59 years old followed for ca. 6 yrs, after adjustment for total energy intake, animal fat was positively associated with the risk of colon cancer (P for trend = 0.01). The relative risk for the highest as compared with the lowest quintile was 1.89 (95% CI, 1.13 to 3.15). The relative risk of colon cancer in women who ate beef, pork, or lamb as a main dish every day was 2.49 (95% CI 1.24 to 5.03), as compared with those reporting consumption less than once a month. Together with the large study in men by Giovanucci *et al.*, (1994), these prospective data provided strong evidence for the hypothesis that a high intake of animal fat from meat increases the risk of colon cancer.

However, three large recent prospective studies have failed to find an association between dietary fat intake and/or meat consumption and cancer risk (Gaard *et al.*, 1996; Veiered *et al.*, 1997; Kato *et al.*, 1997). In the study by Kato *et al.* (1997) of 14,727 women aged 34-65 years who were followed on average for 7 yrs, there was no association of colorectal cancer risk with intakes of total fat, saturated fat or calories, or red meat. In a study in 50,535 Norwegian men and women aged 20-54, with a mean follow-up of 11.4 years, the frequency of consuming meat, including meat stews and roasted meat, was not associated with risk of colon cancer. No trends in relative risks of colon cancer were found with energy-adjusted intake of total fat or saturated fat (Gaard *et al.*, 1996).

Recently, two extensive reviews summarised the existing epidemiologic data on meat and cancer (Baghurst et al., 1997; American Institute for Cancer Research, 1997). In the review by Baghurst et al. (1997) citing over 200 references, the authors conclude the association between meat consumption and cancer risk is largely unproven. They suggest that any true effect of meat consumption alone is likely to be minor, and may be an artefact of a decreased consumption of fruit, vegetables and cereals by high meat consumers. The authors also review potential mechanisms by which meat could be associated with increased cancer risk. These include: 1) high fat intake increasing bile acid secretion and subsequent conversion to mutagens, 2) increased production of potentially toxic fat oxidation products; 3) the production of heterocyclic amines (HA) and polyaromatic hydrocarbons (PAH)) in meat by certain cooking techniques. Regarding the risk from HA and PAH, the authors emphasise that the usual level of exposure remains unclear, as well as the actual risk associated with increasing exposure (see further discussion below).

The American Institute for Cancer Research (1997) has recently published an extensive report (including >3000 references) on diet and cancer in an attempt to establish quantitative recommendations on dietary changes to reduce cancer risk. It includes discussions of cancer by site, dietary components, such as fat and protein, as well as specific foods (meat, eggs, fruits, vegetables, etc.). The authors concluded there is no 'convincing' evidence for an association of meat consumption with cancer at any site, although the data suggest a 'possible' link between red meat and cancers at several sites. In contrast, the evidence is 'convincing' that risk of cancer is influenced by dietary intake of fruits and vegetables (AICR, 1997).

These two comprehensive reviews indicate although the current evidence is suggestive of a link between red meat and cancer risk, overall the data are far from conclusive, mainly due to the difficulty of excluding other confounding factors.

Dietary Patterns

Dietary patterns and dietary variety appear to influence cancer risk associated with meat consumption. A recent study by Singh et al. (1998) indicated increased cancer risk with higher meat intake, but the risk associated with red meat was reduced when legumes were added to the diet. In a recent case-control study, there were significant associations with risk of adenomatous polyps by sigmoidoscopy and higher intake of red meat, animal fat, and saturated fat, while grains, vegetables, and fruits appeared to be protective (Haile et al., 1997). In contrast, in another case-control study (Franceschi et al., 1997), red meat consumption was not associated with colorectal cancer risk, whereas intake of bread, pasta, and refined sugar was associated with increased risk. Several recent studies looking at meat-associated factors and cancer risk have suggested that risk may only be associated with very high intakes. At the highest intakes, saturated fat (Bairati et al., 1998), foods high in heterocyclic amines (Wu et al., 1997) and red meat (Fernandez et al., 1997) were all associated with cancer risk, but no increased risk was seen at low or moderate levels of meat intake. Studies have also examined diet diversity and eating habits and their association with risk of colon cancer (Slattery et al., 1997; Slattery et al., 1998). In a study which categorised subjects by general eating patterns, people who substituted poultry for red meat, whole grains for refined grains, and ate more fruit had a nonsignificant reduction in risk for colon cancer (Slattery et al., 1998).

Gender Differences

Two recent studies suggest gender differences may play a role. In a population-based case-control study of diet composition and risk of colon cancer by Slattery *et al.* (1997) in 1,993 cases and 2,410 controls, higher intake of meats, poultry, fish, and eggs was not associated with risk among women, but was associated with a 50 percent increase in risk among men (95 %CI ,1.1 to 2.0; P trend = 0.01). Similarly, a case-control study in Hawaii reported increased risk for colorectal cancer in men with the high intakes of red meat, but not in women (LeMaarchand *et al.*, 1997).

Heterocyclic Amines

Although some studies have linked meat intake with increased risk of colon cancer, the potential mechanisms for this effect remain unclear. Carefully-controlled experiments in rats exposed to known carcinogens have consistently shown n-6 polyunsaturated fats to be cancer promoters, while feeding even high amounts of saturated animal fat has no promoting effect (Weisburger, 1994). Also, the epidemiologic evidence does not generally support the hypothesis that total fat, or saturated fat, increases risk of colon cancer (Giovannucci & Goldin, 1997). Because of this, other meat components besides fat have been examined as possible risk factors.

Heterocyclic amines (HAs), produced on the surface of meats during roasting or braising, are potent mutagens (Giovanucci *et al.*, 1997). Although they have been implicated as a causal factor in links between meat intake and cancer (Ward *et al.*, 1997; De Stefani *et al.*, 1997; Probst-Hensch *et al.*, 1997), not all studies agree (Ambrosone *et al.*, 1998), and many important questions remain unanswered. HA content of meat can vary widely by cooking method, degree of doneness and use of gravy (Sinha *et al.*, 1998; Skog *et al.*, 1997; Knize *et al.*, 1998). Varying food preparation methods may reduce the risk of certain HAs but not others (Salmon *et al.*; 1997). In addition, it appears that there is a wide range of HA exposure in humans, and typical exposure levels may be much lower than previously thought (Augustesson *et al.*, 1997; Byrne *et al.*, 1998).

N-nitroso compounds

Nitrate and nitrite are added to certain cured meat products and significant levels of nitrosodimetylamine have been found in various processed (salted, cured) meats (Gangoli *et al.*, 1994). Because N-nitroso compounds (NOC) are strongly carcinogenic in animal studies, dietary recommendations caution against the frequent consumption of cured meats. Responding to these concerns, the meat industry has reduced the amount of nitrate added to processed meats to low levels so that in most cases only small amounts of NOC are produced from these additives. Their carcinogenic potential appears to be minor (Higgs & Pratt, 1998). In a recent comprehensive review of NOCs and cancer (Table 1), the authors concluded that, although animal studies have shown NOCs to be potent carcinogens, in humans there is no firm epidemiological evidence linking stomach or brain cancers to dietary intake of nitrate, nitrite and N-nitroso compounds (Eichholzer & Gutzwiller, 1998). This is supported by a recent study showing no association between cured meat consumption and brain cancer (Murphy *et al.*, 1998).

Table 1.   Large Case-Control Studies of Dietary NOC Intake and Stomach Cancer.

Reference and No. of Cases	Dietary Variable	Association	Odds Ratio (95% CI)
Risch et al. (n=246)	Dietary N-Nitroso-dimethylamine	Not significant	0.94 (0.14-6.13)
Gonzalez *et al.* (n=354)	Dietary Nitrosamine	Increased risk	2.09 (p=0.007)
La Vecchia *et al.*(n=746)	Dietary N-Nitroso-dimethylamine	Increased risk	1.37 (1.10-1.70)
Correa *et al.* (n=391)	Sausage and cured meats	Increased risk	2.32 (1.10-4.87)
Boeing *et al.*(n=143)	Dietary Nitrate (incl. smoked meats)	Not significant	1.26 (0.59-2.70)

For references, see text.

Conclusions

Despite public perceptions that consumption of red meat increases risk for cancer, scientific reviews of the epidemiologic data and results from most recent studies do not firmly support this contention. The inconsistencies in the available data suggest that any true effect of meat is likely to be small. However, to better answer these questions, there is clearly a need for further well-controlled prospective trials and for better biomarkers of risk. (Kaaks, 1996; Fair *et al.*, 1997). Until then, current healthy trends in meat production and consumption should be encouraged. Fat intake from meat is decreasing in many countries; for example, in the United Kingdom fat intake from meats fell 25% from 1982 to 1992 (Table 2) (Ulbricht; 1995).

Table 2.   Total dietary fat per person in the United Kingdom, contributed by selected meat groups in 1982 and 1992.

Food group	Dietary fat (kg per person per year) 1982	1992
Beef and veal	2.09	1.93
Lamb	0.66	0.51
Pork and products	6.54	4.36
Poultry	9.29	6.80
Total	18.58	13.60

From Ulbricht (1995) *Fat in the Food Chain.*

Meat producers have responded to concerns over fat intake by selective breeding and feeding practices to favour leaner production. Changes in butchery techniques have further reduced the fat content of many cuts of meat. For the consumer, the current recommendation from the American Institute of Cancer Research appears prudent. It states: "Limit intake of red meat to less than 80 grams (3 ounces) daily. It is preferable to choose fish, poultry and meat from non-domesticated animals in place of red meat."(AICR, 1997). Healthy cooking and preparation methods need to be stressed, as well as the importance of dietary variety, balance and moderation.

References

Ambrosone, C.B., J.L. Freudenheim, R. Sinha, S. Graham, J.R. Marshall, J.E. Vena, R. Laughlin, T. Nemoto & P.G. Shields, 1998. P.G. Breast cancer risk, meat consumption and N acetyltransferase (NAT2) genetic polymorphisms. Int. J. Cancer 75: 825-830

American Cancer Society. Cancer Facts & Figures, 1998. Atlanta, GA; American Cancer Society

American Institute for Cancer Research, 1997. Food, Nutrition and the Prevention of Cancer: a global perspective. Washington, DC: Am. Instit. Cancer Res.

Augustesson, K., K. Skog, M. Jagerstad & G. Steineck, 1997. Assessment of the human exposure to heterocyclic amines. Carcinogenesis 18:1931-1935

Baghurst, P., S. Record & J. Syrette, 1997. Does red meat cause cancer? Aust. J. Nutr. Diet. 54 (4): 1-44

Bairati, I., F. Meyer, Y. Fradet & L. Moore, 1998. Dietary fat and advanced prostate cancer. J. Urology 159: 1271-1275

Boeing, H., R. Frentzel-Beyme, M. Berger, M. et al., 1991. Case control study of stomach cancer in Germany. Int. J. Cancer 47: 858-864

Byrne, C., R. Sinha, E.A. Platz, E.A., E. Giovannucci, G.A. Colditz, D.J. Hunter, F.E. Speizer & W.C. Willett, 1998. Predictors of dietary heterocyclic amine intake in three prospective cohorts. Cancer Epidemiol., Biomarkers & Prev. 7: 523-529

Correa P., E. Fontham, L.W. Pickle L.W. et al., 1985. Dietary determinants of gastric cancer in South Louisiana inhabitants. J. Natl. Cancer Inst. 75: 645-54

De Stefani, E., A. Ronco, M. Mendilaharsu, M. Guidobono & H. Deneo-Pellegrini, 1997 Meat intake, heterocyclic amines, and risk of breast cancer: a case-control study in Uruguay. Cancer Epidemiol. Biomarkers Prev. 6: 573-581

Eichholzer, M., & F. Gutzwiller, 1998. Dietary nitrates, nitrites, and N-nitroso compounds and cancer risk: A review of the epidemiologic evidence. Nutr. Rev. 56(4): 95-105

Fair, W.R., N.E. Fleshner, W. Heston, 1997. Cancer of the prostate: A nutritional disease? Urology 50: 840-848

Fernandez, E., C. La Vecchia, B. D'Avanzo, E. Negri & S. Franceschi, 1997. S. Risk factors for colorectal cancer in subjects with family history of the disease. Br. J. Cancer 75(9): 1381-1384

Franceschi, S., A. Favero, C. La Vecchia, E. Negri, E. Conti, M. Montella, A. Giacosa, O. Nanni, & A. Decarli, 1997. Food groups and risk of colorectal cancer in Italy. Int. J. Cancer 72: 56-61

Friedman, G.D., 1994. Primer of Epidemiology. Fourth Ed. New York: McGraw-Hill, Inc.

Gaard, M., S. Tretli & E.B. Loken, 1996. Dietary factors and risk of colon cancer: a prospective study of 50,535 young Norwegian men and women. Eur. J. Cancer Prev. 5(6): 445-454

Gangioli, S.D., P. A. van den Brandt V.J. Feron V.J. et al., 1994. Nitrate, nitrite and N-nitroso compounds. Eur. J. Pharmacol. 292: 1-38

Giovannucci, E. & B. Goldin, 1997. The role of fat, fatty acids, and total energy intake in the etiology of human colon cancer. Am. J. Clin. Nutr. 66 (6 Suppl.): 1564S-1571S

Giovannucci, E., E.B. Rimm, M.J. Stampfer et al., 1994. Intake of fat, meat and fiber in relation to risk of colon cancer in men. Cancer Res 54: 2390-2397

Gonzalez C.A., E. Riboli, J. Badosa et al. 1994. Nutritional factors and gastric cancer in Spain. Am. J. Epidemiol. 139: 466-473

Haile, R.W., J.S. Witte, M.P. Longnecker, N. Probst-Hensch, M.-J. Chen, J. Harper, H.D. Frankl & E.R. Lee, 1997. A sigmoidoscopy-based case-control study of polyps: macronutrients, fiber and meat consumption. Int. J. Cancer 73: 497-502

Higgs, J., & J. Pratt, 1998. Meat, Poultry and Meat Products. In: Encyclopaedia of Human Nutrition. Sadler, M.J, Strain, J.J., Caballero, B. (editors). Academic press, San Diego, 1272-1282

Kaaks, R., 1996. The epidemiology of diet and colorectal cancer: review and perspectives for future research using biological markers. Ann. Ist. Super Sanita. 32(1): 111-121

Kato, I., A. Akhmedkhanov, K. Koenig, P.G. Toniolo, R.E. Shore & E. Riboli, 1997. Prospective study of diet and female colorectal cancer: the New York University Women's Health Study. Nutr. Cancer 28: 276-281

Knize, M.G., R. Sinha, E.D. Brown, C.P. Salmon, O.A. Levander, J.S. Felton & N. Rothman, 1998. Heterocyclic amine content in restaurant-cooked hamburgers, steaks, ribs, and chicken. J.Agric.Food Chem., Nov.

La Vecchia, C., B. D'Avanzo, L. Airoldi L., et al., 1995. Nitrosamine intake and gastric cancer risk. Eur. J. cancer Prev. 4: 469-74

Langseth, L., 1996. Nutritional Epidemiology: Possibilities and Limitations. Washington, DC: International Life Sciences Institute

Le Maarchand, L., L.R. Wilkens, J.H. Hankin, L.N. Kolonel & L.-C. Lyu, 1997. A case-control study of diet and colorectal cancer in a multiethnic population in Hawaii (United States): lipids and foods of animal origin. Cancer Causes and Control 8: 637-648

Murphy, R.S., C.J. Sadler & W.J. Blot, 1998. Trends in cured meat consumption in relation to childhood and adult brain cancer in the United States. Food Control 9: 299

Probst-Hensch, N.M., R. Sinha, M.P. Longnecker, J.S. Witte, S. A. Ingles, H.D. Frankl, E.R. Lee, & R.W. Haile, 1997. Meat preparation and colorectal adenomas in a large sigmoidoscopy-based case-control study in California (United States). Cancer Causes and Control 8: 175-183

Risch, H.A., M. Jain, N.W. Choi et al. 1985. Dietary factors and the incidence of cancer of the stomach. Am. J. Epidemiol. 122: 947-59

Salmon, C.P., M.G. Knize & J.S. Felton, 1997. Effect of marinating on heterocyclic amine carcinogen formation in grilled chicken. Food Chem. Toxicol. 35: 433-441

Singh, P.N. & G.E. Fraser, 1998. Dietary risk factors for colon cancer in a low-risk population. Am.J.Epidemiol. 148: 761-774

Sinha, R., N. Rothman, C.P. Salmon, M.G. Knize, E.D. Brown, C.A. Swanson, D. Rhodes, S. Rossi, J.S. Felton & O.A. Levander, 1998. Heterocyclic amine content in beef cooked by different methods to varying degrees of doneness and gravy made from meat drippings. Food Chem. Toxicol. 36: 279-287

Skog, K., K. Augustsson, G. Steineck, M. Stenberg & M. Jagerstad, 1997. Polar and nonpolar heterocyclic amines in cooked fish and meat products and their corresponding pan residues. Food Chem. Toxicol. 35: 555-565

Slattery, M.L., T.D. Berry, J. Potter & B. Caan, 1997. Diet diversity, diet composition, and risk of colon cancer (United States). Cancer Causes and Control 8: 872-882

Slattery, M.L., K.M. Boucher, B.J. Caan, J.D. Potter & K.-N. Ma, 1998. Eating patterns and risk of colon cancer. Am. J. Epidemiol. 148: 4-16

Ulbricht, T.L.V., 1995. Fat in the Food Chain. A report to the Ministry of Agriculture, Fisheries and Food. MAFF, London

Veiered, M.B., P. Laake & D. S. Thelle, 1997. Dietary fat intake and risk of prostate cancer: A prospective study of 25,708 Norwegian men. Int. J. Cancer 73: 634-638

Ward, M.H., R. Sinha, E.F. Heineman, N. Rothman, R. Markin, D.D. Weisenburger, P. Correa & S.H. Zahm, 1997. Risk of adenocarcinoma of the stomach and esophagus with meat cooking method and doneness preference. Int. J. Cancer 71: 14-19

Weisburger J.H., 1994. Fleischkonsum, Karzinogenese und Herz-Kreislauferkrankungen. In: Fleisch in der Ernährung, Kluthe R. &. H. Kaspar (editors), Thieme, Stuttgart, 64-74

Willett W.C., M.J. Stampfer, G.A. Colditz, B.A. Rosner & F.E. Speizer, 1990. Relation of meat, fat, and fiber intake to the risk of colon cancer in a prospective study among women. N Engl J Med, Dec 13; 323:1664-1672

Wu, A.H., M.C. Yu & T.M. Mack, 1997. Smoking, alcohol use, dietary factors and risk of small intestinal adenocarcinoma. Int. J. Cancer 70: 512-517

Who eats meat: factors effecting pork consumption in Europe and the United States

W. Jamison

Worcester Polytechnic Institute, Interdisciplinary and Global Studies Division, Worcester, Massachusetts, 01609-2280

Summary

The pork industry in Europe and the United States is characterised by excellent parameters of production and quality. Indeed, the efforts of researchers and industry have produced pork products that are lean, nutritious, cost effective and competitively priced, and increasing levels of efficiency marks pork production. Furthermore, the current price of pork to consumers is extremely competitive, particularly in the United States where select cuts of pork are priced equivalently to chicken. Hence, an increase in overall per capita pork consumption could be expected. However, that is not the case. Pork consumption in Europe and the U.S. has declined or remained stagnant for the last several years. This raises an interesting question: why, in the face of favourable production parameters, excellent nutritional and price value, has pork consumption stagnated and declined? This paper examines this question and forwards the hypothesis that, in addition to market depressing factors such as general negative impressions of meat and market saturation, consumers in the U.S. and Europe are increasingly making purchasing decisions based upon factors other than cost, nutritional value and availability. Indeed, some consumers consider alternative criteria when deciding whether to buy pork. Whereas industry and academia have focused on extrinsic values like increased production efficiency, lower retail cost, lower fat and higher nutritive value in their efforts to increase pork consumption, consumers in modern societies no longer respond solely to those criteria. Indeed, consumers increasingly rely upon intrinsic factors when making purchasing decisions.

Keywords: Intrinsic, Extrinsic, Pork, Consumers, Values

Introduction

Beginning with the Industrial Revolution, radical changes in agriculture have transformed animal production from local subsistence activities to specialised, industrialised, single animal operations that maximise economies of size and scale. Pork production since the 1950's has been characterised by this astonishing technological change. Much of this change has been motivated by simple market economics, which posit that the product that most completely matches consumer demand gains a competitive advantage. Thus, intense efforts in pork production have focused upon growing an animal that is both profitable and efficient. Indeed, recent developments in agricultural science such as genetic engineering promise to increase the efficiency of production parameters even further. Likewise, most of the developments in animal agriculture over the last 20 years have been driven by economic rationalism, whereby consumers make decisions based upon rational choices in an open marketplace between competing meat products. In response, animal agriculture has pushed incessantly to lower costs, increase efficiency, and provide a safe, healthy, plentiful, low cost meat product to consumers. The presupposition is that the meat product that best provides these values to the consumer will have the competitive advantage. While certainly true to some extent, this approach has encountered difficulties in recent years. More important, while advances in agricultural technology have greatly increased pork production parameters, a concurrent and far more important revolution has transformed the way information about pork production is assimilated by the public. Indeed, the information revolution has had far-reaching and poorly understood influences on animal production.

This paper examines the factors that influence pork consumption in Europe and the U.S., and focuses primarily on the changes wrought by the information revolution. It begins by reviewing traditional parameters that have influenced consumers since 1945. It then reviews current trends in pork consumption and discovers that, in many countries, pork consumption has stagnated or declined over the last several years. This stagnation comes in the face of production parameters that should belie a gain in market share: pork is competitively priced and is the leanest and healthiest in its history. Yet, pork consumption is stagnant. Why? The paper argues that the industry and academia have been responding to extrinsic values, seeking to fulfil demands of rational consumers who make rational cost-benefit equations when they buy meat products. Meanwhile, modern societies have undergone fundamental changes that greatly effect the way consumers assimilate knowledge and act upon their needs. Whereas the pork industry has emphasised extrinsic values, consumers increasingly make decisions based upon intrinsic values. More important, the ability of the pork industry to influence pork consumption by providing supportive extrinsic data is limited by these fundamental changes.

Likewise, the paper presents evidence that supports the hypothesis that consumers in the U.S. and Europe increasingly rely upon factors to make their decisions that are unknown to producers. Furthermore, evidence is presented that argues that consumers are decreasingly responsive to the traditional methods of information dissemination employed by the pork industry. Lastly, the paper argues that pork consumption is influenced by a myriad of poorly understood factors, including information overload, distrust of information sources, and distrust of pork production and its supporting institutions.

Extrinsic v. Intrinsic Values

Pork producers have operated under the rubric of economic rationalism since at least the 1950's. In approaching the industrial production of pigs, farmers, processors, retailers and universities have assumed that consumers make rational decisions based upon a cost : benefit analysis of the meat product. The presupposition behind modern pork production implies that meat products whose benefits outweigh their costs will be purchased. Indeed, the assumption has been that the meat product that best provides these benefits will out-compete the others for competitive market share. Hence, much emphasis has been placed upon identifying and providing these consumer benefits. The benefits include excellent taste and food quality, low fat, competitive price, attractive appearance, and outstanding health and food safety attributes. Another way to view these benefits are as values with which the consumer makes decisions.

It almost goes without saying that the pork industry has been effective in providing these values. As the industry provided these values, pork consumption grew steadily since the 1940's. Indeed, worldwide pork consumption increased steadily from 1949 through 1990. Yet, beginning in the late 1980's, pork consumption patterns in western countries have changed. This trend is mirrored in many nations in Europe as well as the United States: pork consumption increased steadily until stagnation and decline began in the 1990's (USDA, 1998). This is interesting in that pork production has continued to innovate, becoming more and more efficient and increasingly adept at responding to parameters of production. Pork is lean, cost effective, healthy, and appealing, as judged by consumption parameters of the 1980's. Yet, consumption patterns have changed in both the U.S. and Europe. Why is this so, and what is behind the shift in pork consumption? This paper argues that consumers in modern, affluent societies are increasingly using alternative values in making their decisions. Instead of using extrinsic values such as rational data on nutrition and production, they are increasingly using intrinsic values that bear little relation to, and are often divorced from, those values being provided by pork producers.

Extrinsic values are defined as external information sources which attempt to influence and motivate behavior, and are typified by appeals to rational decision processes. These values include information, knowledge, data, and are generally available throughout society, e.g. "one size fits all." Indeed, extrinsic values tend to deal with cost, health, safety, risk, and benefit to society. Whereas extrinsic values emphasise general values that apply to most people, intrinsic values are highly individualistic and self-actualised. Intrinsic values are defined as internal information, feeling, or beliefs that act as presuppositions and normative beliefs (Berger & Luckman, 1966; Durkheim, 1973; Berger *et al.*, 1974). In other words, intrinsic values are the underlying values through which a consumer filters all other information and decides which information to believe. Intrinsic values are often sub-conscious; the consumer rarely is aware of the assumptions that colour his decisions. Intrinsic values include feelings, emotions, and personality attributes, as well as beliefs, attitudes, and moral and ethical opinions. Intrinsic values reveal whether something is right or wrong, while extrinsic values weigh the relative, rational merits of something (Reiss & Straughan, 1996). A question of whether pork production is wrong is an intrinsic evaluation, whereas the question of whether pork is economically competitive with other meat products is an extrinsic evaluation. In other words, extrinsic values are openly available to all consumers, are general data concerning the benefits or costs of something, and deal with rational, conscious decision processes. In contrast, intrinsic values are internalised, highly individualistic presuppositions and normative ideals that are often sub-conscious and hard to identify.

What does this have to do with pork consumption? Pork producers, animal scientists and retailers of pork products have identified many of the extrinsic factors motivating pork consumption. Furthermore, they have been largely successful in providing these values to consumers. Nevertheless, intrinsic values are increasingly prevalent in buying decisions, often times to the exclusion of extrinsic values. Likewise, self-actualisation and rapidly increasing numbers of intrinsic values typify modern affluent societies. In other words, countries that have a high standard of living also tend to have a plethora of belief systems wherein values are highly individualised and atomised.

This means that the pork industry, which has implicitly accepted extrinsic values as its measurements of production parameters, is ill equipped to address the intrinsic values with which modern consumers increasingly make decisions. While excelling at providing low cost, high quality meat in a competitive environment (extrinsic values), the industry has only begun to understand the role of intrinsic values in pork consumption.

Evidence

Certainly, many consumers continue to make decisions based upon extrinsic values. Indeed, if confronted by competitive meat products, extrinsic factors often sway the buying decision. That is evidence by the effect that retail discount sales have upon meat consumption: the lower the price, the greater the consumption. However, intrinsic values are clearly a factor in buying decisions. Whether the BSE scare in Britain or the tainted meat dilemma in Belgium, weather animal protection regulations or an emphasis on environmental awareness, clearly consumers are increasingly interested in values that appear to be separate from the data supporting meat consumption. In other words, there is little scientific evidence supporting the risk of BSE in pork, yet consumers were scared of pork nonetheless. There is little scientific evidence that stringent animal welfare regulations actually makes a pig happier, yet consumers in Europe demand that animals be treated "humanely." To put it another way, the extrinsic evidence behind each of these consumer issues was minimal, yet the issue was highly emotional, therefore visible and political.

A study conducted by a team of European social scientists sheds light on this phenomenon. A very large sample of consumers throughout the EU was evaluated concerning attitudes about biotechnology and trust in information sources concerning food production and animal research. The study found that consumers placed little trust in universities, farmers, or industry officials regarding food production and biotechnology. Furthermore, the study indicated that consumers placed very little trust in any entity that appeared to have a vested interest in the outcome. Instead, the study indicated that consumers place the highest trust in environmental and consumer organisations concerning food production, and place the most trust in the medical profession for information about biotechnology. Likewise, the study indicated that consumers are suspicious of the information sources and experts that should know the most about the topic: farmers, universities and industry officials were viewed with suspicion concerning food production (Bauer, et al., 1997). Similarly, the study indicated that many public policy debates surrounding food and biotechnology are moral and political instead of being based upon economic, cost : benefit, or risk assessment evaluations. In other words, the study supports the finding that consumers are increasingly relying upon intrinsic values to makes decisions rather than extrinsic values provided by outside information sources and experts.

The literature indicates that consumers in modern economies are increasingly prone to make decisions based upon intrinsic values. At least three factors support this contention. First, evidence indicates that affluence closely correlates with meat consumption. As a society increases in affluence, its meat consumption increases. Based upon various factors, including adequate disposable income and a perception of meat as a luxury, consumers buy more meat as their income increases. However, this holds true only to a point of diminishing returns, at which point the affluence of a culture allows citizens to indulge in self-actualisation and alternative values, often in opposition to the production and consumption of meat (Rieder, 1997). In other words, in Europe and the rest of the westernised world, ethical vegetarianism is typical only in countries that have reached a comfortable level of affluence. While this hypothesis is useful, it fails to adequately explain the phenomenon of intrinsic values in western countries. While it is true that meat consumption increases as affluence increases, there is ample evidence that affluence need not determine attitudes towards meat. A case in point is Japan and many affluent Asian countries, where not only has meat consumption remained relatively stable as affluence has increased, but people eat any animal without compulsion or moral protest.

Another factor that helps explain decision making based upon intrinsic values is the increasing complexity of modern societies. As society becomes increasingly complex, people are decreasingly able to understand complex phenomena. For example, only a monopsony of elites understands the genetic manipulation of pigs, while the vast majority of citizens cannot fathom the scientific data. Thus, people tend to be sceptical, suspicious and scared of what they cannot understand. In turn, they retreat from the debate and turn away from the rational evaluation of extrinsic values in favour of trust in either their own intrinsic values or information that supports them (Beck, 1992; Bauer, 1995). What this means is that complex societies are marked by increasingly complex debates where public policy is dominated by highly esoteric and specialised data, and that these complex societies are characterised by citizens who feel alienated and aloof from the rational, scientific discourse. In the case of pig production, as ethologists, physiologists, and endocrinologists debate the merits of animal welfare, the consumer resists supportive extrinsic values presented by the scientists. In other words, as complexity increases, people shut down their rational decision processes based upon extrinsic values, opting in favour to depend upon their own values and beliefs to make decisions (Giddens, 1990). This has the effect of insulating them from the data and places an emphasis on whether something "feels" right (Coghlan, 1993).

A third factor is the immense amount of information available to a consumer, e.g. the immense amount of extrinsic values. In the information age a consumer is confronted with an enormous amount of information that is difficult to understand, is often contradictory or false, and is overwhelming in its volume and magnitude. In response, the consumer retreats from the public forum, forgoing extrinsic information from unknown or untrustworthy sources in favour of intrinsic values and information that confirms and validates those values from trusted sources (Beck, 1992). Thus, genetic manipulation and cloning in pigs is marked by such a cacophony of information. A panoply of experts provides extrinsic data based upon risk analysis and cost : benefit analysis to consumers who are neither capable nor willing to evaluate the data. In addition, contradictory experts making contradictory claims characterise the public debate, as an expert from Greenpeace contradicts an expert from the Swiss Federal Institute of Technology. Furthermore, information is available from multiple media and the Internet and is not subject to verification of its accuracy. This results in a general distrust of all extrinsic information, and a retreat from the public forum in favour of reliance upon intrinsic values and trusted, similar thinking individuals.

Conclusion

This paper has not been intended to be an exhaustive over-view of the literature on consumer psychology or the sociology of markets and purchasing decisions, nor has it been intended to provide an definitive definition of extrinsic and intrinsic values. Certainly the phenomenon of extrinsic and intrinsic values is far more complex than stated here, and little justice has been done to the interchange and synthesis of the two values in the decision making process. Instead, a rudimentary hypothesis has been forward that, at least, pork consumption is increasingly influenced by factors that were previously unidentified and undisclosed in the pork industry. Indeed, as the controversies and issues surrounding the pork industry in the last few years have indicated, many public debates about pork consumption and therefore private buying decisions, are moral and political rather than scientific. Certainly, pork producers, animal scientists and industry personnel have been highly successful in the past in providing outstanding products at competitive values. The past fifty years bear witness to the ability of the pork industry to match consumer demand. In effect, the industry has been adept at identifying and fulfilling the extrinsic values of European and American society. This success has predicated upon the presupposition that people make rational decision based upon extrinsic values that are carefully weighed in a cost : benefit analysis. The meat product that best supplies the most benefits for the lowest costs will win the competitive market for meat consumption. This has certainly been the case with pork consumption between 1945 and 1985. Most people affiliate with the pork industry would agree with these factors: consumers want a pork product that tastes good, is healthy and lean, is safe and pure, and is inexpensive. Furthermore, they might even agree that consumers now pay attention to such things as origin of the product, and the animal welfare and environmental standards under which it was produced.

Nevertheless, even with the industry's success at providing extrinsic values, pork consumption has stagnated or declined in Europe and the U.S. This paper argues that a principal factor of this decline may be the increased reliance of consumers upon intrinsic values in their decisions. Albeit, extrinsic values such as cost and nutritive value are critical to success of the pork industry, the author believes that intrinsic values now filter and even supercede those extrinsic parameters. In other words, the pork industry must not only continue to provide inexpensive, high quality pork, it must also begin to account for the intrinsic values of its consumers.

Consumers in affluent, modern countries have the luxury of seeking self-actualisation in the form of ever-increasing and divergent intrinsic values. Whether as an animal rights activist or ethical vegetarian, whether as an environmental advocate or an opponent of "unnatural" production methods, modern society affords the ability to express intrinsic values in myriad and ever changing ways. At the same time, extrinsic values that have traditionally vied for the consumer's attention are increasingly complex, esoteric and undecipherable. The same vegetarian who places emphasis on the ethics of eating is simply incapable of truly understanding the scientific data concerning cloning pigs. In addition, the consumer in modern societies is confronted with information overload, data that is overwhelming in its volume and intensity, and is often contradictory and false. Thus, the ethical vegetarian has the ability to actualise her intrinsic values, the inability to understand and evaluate the extrinsic values that seek to influence her decision process, and is overwhelmed by the extrinsic information that confronts her. Thus, she increasingly makes decisions based upon intrinsic evaluations.

What does this mean for the pork industry? In the past, whereas providing extrinsic values was sufficient for success, the modern consumer makes decisions based upon a panoply of values. These values are not only based upon information and knowledge, but also upon feelings and emotions. Indeed, it is argued that modern consumers cannot understand the science of pork production, they have the ability and will to self-actualise alternative intrinsic values, and they are incapable and unwilling to sift through the information that is available to help them make rational decisions based upon extrinsic data. Instead, they trust the "self", reject outside information and organisations that disconfirm their intrinsic values, and seek out and place trust in information that supports and validates their beliefs. This means that the pork industry and the scientists who support it should expect to be increasingly marginalised in the debate over pork production and agricultural policy. Likewise, it means that consumers will be intolerant of the pork industry as it attempts to implement new technologies that seek to meet consumer extrinsic demands while offending their intrinsic values. It will become increasingly problematic to provide taste, health, cost, and nutritional advantages to consumers without offending their intrinsic values. Similarly, when those values are offended and public debate rages about the new technologies, the pork industry will be unable to have itself heard as it tries to provide data to shape and inform the debate. In other words, the pork industry is caught between a rock and a hard place: it must continue to emphasise the production parameters needed to meet consumer demand, while anticipating and responding to intrinsic demands that contradict those parameters. In response, the industry can only begin to research, identify and adapt to those intrinsic values in much the same way as it did to the extrinsic demands of consumers earlier in the century.

References

Bauer, M. (editor), 1995. Resistance to New Technology: Nuclear Power, Information Technology, Biotechnology. Cambridge University Press, London

Bauer, M., J. Durant & G. Gaskell, 1997. Europe Ambivalent on Biotechnology. Nature 26 June: 845-847

Beck, U., 1992. Risk Society: Towards a New Modernity. Sage, London.

Berger, P., B. Berger & H. Kellner, 1974. The Homeless Mind: Modernization and Consciousness. Vintage, Washington, D.C., 258pp.

Berger, P., & T. Luckman, 1966. The Social Construction of Reality: A Treatise in the Sociology of Knowledge. Anchor, New York, 219 pp.

Coghlan, A, 19 June, 1993. Gene Industry Fails to Win Hearts and Minds. New Scientist 19 June, 1993: 4

Durkheim, E., 1973. On Morality and Society. University of Chicago Press, Chicago, 244pp.

Giddens, A., 1990. The Consequences of Modernity. Cambridge University Press, London

Reiss, M., & R. Straughan, 1996. Improving Nature? Cambridge University Press, London

Rieder, P., 1997. Personal Communication

United States Department of Agriculture, Foreign Agricultural Service, Commodity and Marketing Division for Dairy, Livestock, and Poultry, 1998. Counselor and attaché reports. GPO, Washington, D.C.

Genetics and pork quality

Breed effect on meat quality of Belgian Landrace, Duroc and their reciprocal crossbred pigs

G. Michalska, J. Nowachowicz, B. Rak & W. Kapelanski

University of Technology and Agriculture, Mazowiecka 28, 85-084 Bydgoszcz, Poland

Summary

Meat quality traits, such as pH_1, colour lightness and soluble meat protein content, were determined in purebred Belgian Landrace (BL), Duroc (D) and their reciprocal crossbred progeny (BL x D and D x BL). Each group consisted of 30 gilts slaughtered at 185 days of age. The heterosis effect for pH_1 was minimal (1.15 for BL x D and 0.82% for D x BL) but for colour lightness was distinct (-13.11 for BL x D and -8.76% for D x BL), and was medium (4.35%) for soluble protein content in meat in both crossbred groups of pigs. Meat quality in crossbred pigs was good, such as that of purebred Duroc pigs.

Keywords: Pigs, Crossbreeding, Meat quality, Heterosis

Introduction

The breed effects on meat quality in crossbreeding are not easy to predict and ought to be studied in specific crossing combinations. Since crossbreeding is extensively used in pig production, the knowledge of possible heterosis effects on meat quality is of great interest. The classical view that meat quality traits are additively inherited in breed crosses (i.e. no heterosis) was presented by Sellier (1987). However, when such different breeds with respect to stress resistance as Belgian Landrace and Duroc are crossed, the meat quality from crossbred animals would be better than the average of the parental purebreds.

Numerous studies proved that remarkably meaty breeds, such as Belgian Landrace, are inferior in respect to meat quality (Lampo, 1977; Michalska, 1996; Michalski & Kamyczek, 1985; Michalski & Kamyczek 1988; Ollivier *et al.*, 1976; Sellier & Monin, 1994; Steindel & Kaczmarek, 1980). On the other hand, the Duroc breed is stress resistant and is characterised by good meat quality (Houde *et al.*, 1993; Lo *et al.*, 1992; Sellier & Monin, 1994).

The objective of this study was to compare the meat quality traits of purebred Belgian Landrace and Duroc, as well as their reciprocal crossbred pigs.

Material and methods

The study was carried out on four groups of gilts, being the progeny of purebred Belgian Landrace (BL), Duroc (D) and crossbreds of BL x D and D x BL (sire breed being on first position). Each group consisted of 30 animals reared in the Hybridisation Centre in Pawlowice belonging to National Institute of Animal Production. Pigs were slaughtered at 185 days of age.

Meat quality traits were determined according to methods used in Pig Progeny Testing Stations (Kielanowski *et al.*, 1977). The pH_1 value was measured about 45 minutes after slaughter and 24 hours later the meat colour lightness and meat soluble protein content were determined.

The results were statistically analysed using variance analysis and significance of differences was assessed by the Duncan test (Ruszczyc, 1978). Heterosis effects were calculated according to the equation (Nowicki *et al.*, 1994):

$$E_h = \frac{x_F - x_R}{x_R} \times 100$$

where:

E_h = heterosis effect expressed as %

x_F = value of trait in crossbred animal

x_R = mean value of this trait in parental breeds

Results and discussion

Meat characteristics for purebred and crossbred pigs are presented in Table 1, as well as the heterosis effects for those traits in crossbred pigs. As can be seen, the mean pH_1 value in meat of purebred Belgian Landrace pigs was significantly lower than in Duroc, BL x D and D x BL pigs (P < 0.01). Similarly, meat colour lightness (i.e. paleness), was the highest in these pigs (P < 0.01). It is worthwhile to note that the two crossed groups were better than the purebred in that respect. Soluble meat protein content which is highly correlated with water holding capacity of meat also was the lowest in purebred BL in comparison with purebred Duroc and crossbred BL x D and D x BL pigs (P < 0.01). The above results are in agreement with the data of many authors (Lampo, 1977; Michalska, 1996; Michalski & Kamyczek, 1985; Michalski & Kamyczek 1988; Ollivier *et al.*, 1976; Sellier & Monin, 1994; Steindel & Kaczmarek, 1980).

Table 1. Meat values (x), standard deviations (s) and heterosis effect (Eh) for meat quality traits.

Traits		BL	D	BL x D	D x BL
				Groups	
pH_1	x	5.92[A]	6.25[Ba]	6.15[B]	6.13[Bb]
	s	0.25	0.18	0.20	0.19
	E_h			1.15	0.82
Colour lightness, %	x	28.58[A]	25.75[Ba]	23.60[Bb]	24.78[B]
	s	3.18	1.88	2.12	2.42
	E_h			-13.11	-8.78
Soluble meat protein, %	x	7.60[A]	8.03[B]	8.15[B]	8.15[B]
	s	0.47	0.43	0.39	0.39
	E_h			4.35	4.35

a, b - mean values followed by different letters differ significantly (P < 0.05)
A, B - mean values followed by different letters differ significantly (P < 0.01)

It may be stressed that in the studied crossbred groups the favourable crossing effects that were shown concerned the higher meat quality traits. The crossbred pigs had higher pH_1 value, less pale meat colour and a larger quantity of soluble protein content in meat in comparison with the means of parental breeds.

The heterosis effect appeared to different extents in regard to particular meat characteristics. The lowest heterosis effect was observed in respect to pH_1 value (1.15% in BL x D and 0.82% in D x BL), the highest to colour lightness of meat (-13.11% in BL x D and -8.76% in D x BL) and a medium heterosis effect was found for soluble meat protein content (4.35% in both crossbred groups). A similar result regarding the beneficial heterosis effect on meat colour was stated by Lo *et al.* (1992) in crossbred Duroc and Landrace pigs. According to Sellier (1998) favourable heterosis may be found only in some breed combinations where stress-resistant and stress-susceptible breeds are crossed. This can be interpreted in terms of halothane genotypes and related to the partial recessiveness of the n allele for some PSE meat indicators (Sellier, 1998).

Conclusion

1. Purebred Belgian Landrace pigs were characterised by deteriorated meat quality in comparison with purebred Duroc and crossbred BL x D and D x BL pigs.
2. Meat quality traits in crossbred BL x D and D x BL were at the same level as in the purebred Duroc pigs.
3. Among the studied meat quality characteristics a heterosis effect was highest for meat colour lightness.

References

Houde, A., S.A. Pommier & R. Roy, 1993. Detection of ryanodine receptor mutation associated with malignant hyperthermia in purebred swine populations. J. Anim. Sci. 71: 1414-1418

Kielanowski, J., H. Duniec, T. Kostyra, M. Kotarbinska, F. Maly, Z. Osinska, M. Rozycki & W. Szulc, 1977. [Results of evaluation in pig testing stations of the Institute of Zootechnics in 1976]. PWRiL, Warszawa, 5-28

Lampo, Ph., 1977. [Heritability of stress susceptibility; indicator enzymes and correlations between their blood levels and some economic important characters in Belgian pigs - some preliminary results]. Prz. Nauk. Lit. Zoot. 4 (94): 40-42

Lo, L.L., D.G. McLaren, F.K. McKeith, R.L. Fernando & J. Novakofski, 1992. Genetic analyses of growth, real-time ultrasound, carcass, and pork quality traits in Duroc and Landrace pigs: I. Breed effects. J. Anim. Sci. 70: 2373-2386

Michalska, G., 1996. [Heterosis effects for reproductive performance, growth performance and carcass traits in two-breed reciprocal crosses of Belgian Landrace with Polish Large White and Duroc pigs]. Zesz. Nauk. ATR Bydgoszcz, Hab. Dissertation, No 76

Michalski, Z. & M. Kamyczek, 1985. [The reaction of pure breed and crossbred pigs to halothane test]. Rocz. Nauk. Zoot. 12 (1): 135-141

Michalski, Z. & M. Kamyczek, 1988. [Reaction to halothane anaesthesia in pigs]. Zesz. Probl. Post. Nauk Rol. 335: 263-268

Nowicki, B., E. Pawlina, W. Kurszynski & P. Łos, 1994. Leksykon z zakresu genetyki i hodowli zwierząt. PTZ, Warszawa

Ollivier, L., P. Sellier & G. Monin, 1976. Frequency of the malignant hyperthermia syndrome (MHS) in some French pig populations: preliminary results. III Conf. Production Disease in Farm Animals, Wageningen, Netherlands, 208-210

Ruszczyc, Z., 1978. [Statistical methods in animal experiments]. PWRiL, Warszawa

Sellier P. 1987. Crossbreeding and meat quality in pigs. In Evaluation and Control of Meat Quality in Pigs. Eds. P.V. Tarrant, G. Eikelenboom and G. Monin. Martinus Nijhoff Publ., Dordrecht, The Netherlands, 329-342

Sellier, P., 1998. Major genes and crossbreeding with reference to pork quality. Pol. J. Nutr. Sci. 7/48, No 4(S): 77-89

Sellier, P. & G. Monin, 1994. Genetics of pig meat quality: A review. J. Muscle Foods, 5: 187-219

Steindel, B. & W. Kaczmarek, 1980. [Comparison of meat quality of pure breed Belgian Landrace, Polish Large White, Polish Landrace and Belgian Landrace x Polish Landrace crossbred pigs]. Rocz. Nauk. Zoot. 7(1): 123-130

Genetic and energy effects on pig meat quality

Z. Gajic[1] & V. Isakov[2]

[1] *University of Belgrade, Faculty of Agriculture, 11081 Beograd*
[2] *A.D. "29. Novembar", 24000 Subotica, Yugoslavia*

Summary

Direct and interaction effects of breed (Mangalitsa, Yugoslav Meat Breed and Belgian Landrace) and levels of diet energy (low, medium and high) on meat quality traits were examined in a 3 x 3 factorial experiment. Each of 9 subclasses consisted of 20 pigs all taken off test at 100 kilograms for slaughter.

The three breeds differed regarding some of the meat quality parameters. The Belgian Landrace had significantly lower pH_1 in *M. semimembranosus* (6.12) and nearly the same as the Mangalitsa and Yugoslav Meat Breeds (6.43 and 6.46, respectively). A similar tendency was observed in *M longissimus dorsi*. The meat from Belgian Landrace pigs had a lower water holding capacity, observed as higher wetness value in *M. longissimus dorsi* (4.14 cm^2) than the Mangalitsa (3.75 cm^2) and higher than the Yugoslav Meat Breed (3.56 cm^2). Differences in dry matter and crude protein content between the breeds were significant. Percentage of intramuscular fat in the meat show that the Mangalitsa ranked highest (2.93), followed by the Yugoslav Meat Breed (1.93) and then the Belgian Landrace (1.67). The differences are highly significant.

Keywords: Pigs, Genetic, Energy level, Carcass, Meat quality

Introduction

Strategies and research from production to consumption have been initiated to improve carcass and meat quality. Meat production may be influenced by such variables as genetic, nutritional and environmental factors. Genetic factors play a primary role in the overall variability of technological properties of meat and fat in pigs (Goodwin, 1997). This study aims to determine the effects of genetic and energy levels on carcass composition and meat quality.

Material and methods

In these investigations three breeds of pigs were used: Mangalitsa (fat genotype), Yugoslav Meat Breed (lean genotype) and Belgian Landrace (extra lean genotype). Sixty piglets were taken from each breed and divided into three groups of 20 animals. There were three periods of fattening, one from 20 to 35 kilograms, from 35 to 60 kilograms and from 60 to 100 kilograms of liveweight. The diets used differed with respect to energy content. A balanced 3 x 3 factorial experiment included the three breeds and three levels of energy (low, medium and high - Table 1).

On achieving the liveweight of 100 kilograms, the pigs were slaughtered. After 24 hours of cooling, total dissection of the right side was carried out using the method of Weniger *et al.* (1963).

Table 1. Experimental design.

Body	Digestible energy levels MJ/kg		
weight	M-M	H-L	H-H
20 - 35 kg	13.01	13.37	13.37
35 - 60 kg	13.32	13.63	13.63
60 - 100 kg	13.46	13.03	13.74

Energy levels: M - Medium; L - Low; H - High

At cutting, various meat quality measurements were made. In the dissected ham muscle (*M. semimembranosus*) and *M. longissimus dorsi* pH_1 (after 45 minutes) and pH_{24} (24 hours *post mortem*) were measured with a pH-meter. Water holding capacity in *M. longissimus dorsi* was measured using the Grau-Hamm method. Muscle samples for chemical analysis were taken from *M. longissimus dorsi* between the 13th and 14th ribs. The content of dry matter was determined after drying at 105°C for 4 hours Crude protein was analysed by the Kjeldahl method, using the factor 6.25 when calculating the proportion of crude protein.

Results and discussion

The highest meat content in the cold carcass was found in the M-M treatment and amounted to 47.01% (Table 2). The meat content was somewhat lower in the H-L treatment amounting to 46.81%. The meat level was lowest in the H-H treatment amounting to 46.12%. Based on the statistical analysis the conclusion which tends to emerge is that there is little difference in the meat content between treatments, i.e. between energy levels which were found to be statistically insignificant.

Table 2. The yield of lean meat in the carcass, %.

Breeds	Energy levels			Breed
	M-M	H-L	H-H	mean
Mangalitsa	38.19	37.36	37.17	37.57
Yugoslav Meat Breed	48.08	48.88	47.48	48.15
Belgian Landrace	54.77	54.18	53.70	54.22
Energy levels mean	47.01	46.81	46.12	

The effect of genotype on the content of total meat in the carcass was evident and substantial. The lowest meat content amounting to 37.5% was found in the fat genotype (Mangalitsa). A meat content amounting to 48.15%, (i.e. 28.16% higher than the fat genotype), was found in the lean genotype (Yugoslav Meat Breed). The highest meat content in the carcass amounting to 54.22% was found in the extra lean genotype (Belgian Landrace). The meat content was 44.32% higher in the extra lean genotype than in the fat genotype. All the differences between the meat content in the carcasses of the genotypes tested were statistically highly significant (P>0.01).

The trend of the content of total fatty tissue (Table 3) was opposite to the trend of the meat content in the carcass. The effect of genotype on the content of fatty tissue was more prominent than the effect of energy level. The content of fatty tissue was lowest in the M-M treatment amounting to 37.23%. The fat content was significantly higher in the H-H diet treatment only (38.47%), i.e. at the highest energy level, compared with the M-M and H-L treatment.

Table 3. The yield of fatty tissue in the carcass, %.

| Breeds | Energy levels | | | Breed |
	M-M	H-L	H-H	mean
Mangalitsa	47.56	47.37	48.72	47.88
Yugoslav Meat Breed	35.32	34.29	36.62	35.41
Belgian Landrace	28.82	30.10	30.08	29.67
Energy levels mean	37.23	37.25	38.47	

Based on the results of the effect of genotype on the content of fatty tissue in the carcass, the conclusion which tends to emerge is that the highest content is in the fat genotype. The fat content in this group amounted to 47.88%. In the lean genotype the fat content amounted to 35.41% representing a decrease of approximately 26%. The fat content was lowest in the extra lean genotype. The fat content in the carcass of this group amounted to 29.67%, i.e. it was 38% lower than the fat genotype.

During the last decade, pig production was characterised by the continuous improvement of lean meat and at the same time, a marked reduction of the amount of fat. But it is established that quantity of meat in the carcass and technological quality of meat are negatively correlated at the genetic level. Estimates of genetic correlations between meat quantity and meat quality criteria are between -0.10 and -0.40 (Sellier, 1990).

Measurements of the meat quality indicators are shown in Table 4.

Table 4. Meat quality indicators.

Traits	Mangalitsa	Yugoslav Meat Breed	Belgian Landrace
pH_1 of ham	6.43	6.46	6.12
pH_{24} of ham	5.95	6.16	6.07
pH_1 of MLD	6.42	6.43	6.19
pH_{24} of MLD	5.96	6.13	6.11
Water holding capacity, cm^2	3.75	3.56	4.14
Dry matter content, %	27.53	26.50	25.87
Protein content, %	23.87	24.00	23.17
Intramuscular fat, %	2.93	1.93	1.67

The three breeds differed in regard to some of the meat quality parameters. The Belgian Landrace had significantly lower pH_1 in *M. semimembranosus* (6.12) and nearly the same as the Mangalitsa and Yugoslav Meat Breeds (6.43 and 6.46, respectively). A similar tendency was observed in *M. longissimus dorsi*. The meat from Belgian Landrace pigs had a lower water holding capacity, observed as a higher wetness value in *M. longissimus dorsi* (4.14 cm^2) than Mangalitsa (3.75 cm^2) and higher than the Yugoslav Meat Breed (3.56 cm^2). Differences in dry matter and crude protein content between the breeds were significant. Percent of intramuscular fat in the meat shows that the Mangalitsa ranked highest (2.93), followed by the Yugoslav Meat Breed (1.93) and the Belgian Landrace (1.67). The differences are highly significant.

Conclusion

From these comparative investigations it may be concluded that the Mangalitsa, Yugoslav Meat Breed and Belgian Landrace greatly differ in terms of the meat quality traits studied. No trait was significantly affected by diet energy level. Breed x energy level interactions were not significant for any meat quality traits except dry matter content.

References

Goodwin, R., 1997. Genetic effects on pork quality. The Pork Quality Summit. National Pork Producers Council, Des Moines, IA, 25pp.

Sellier, P., 1990. Genetic control of meat quality. 41st Annual Meeting of EAAP, Toulouse, France, 10pp.

Weniger, J.H., D. Steinhauf & G. Pahl, 1963. Muskeltopographie der Schlachtkörper. Bayer. Landwirtschaftsverlag, München

Genetic parameters for fattening traits in the Belgian Piétrain population

D. Geysen[1], S. Janssens[1] & W. Vandepitte[1]

[1]*Centrum voor Huisdierengenetica en selectie, Department Animal Production, K.U.Leuven, Minderbroederstraat 8, B-3000 Leuven, Belgium.*

Summary

Genetic and phenotypic parameters for fattening traits were estimated for a centrally tested purebred stress susceptible Piétrain population. Data (from 1984 to 1998) consisted of individual records on growth (n = 18,755) and lean meat percent (n = 7,908). Growth was expressed as the number of days needed to grow from 25 kilograms to 100 kilograms liveweight. Lean growth rate (lgr) was calculated from average daily gain and lean meat percentage (lm). All data were precorrected for final and initial weight and sex differences. Reml variance components were estimated using an animal model that included the random direct additive genetic effect of the animal, the random common environmental effect of the litter and the random effect of station-year-season. Resulting heritabilities were 0.33 ± 0.02, 0.27 ± 0.02 and 0.27 ± 0.01 for lm, lgr and days respectively. Genetic correlation between days and lm is moderately unfavourable (0.30 ± 0.03), between days and lgr is highly negative (-0.94 ± 0.01) and is close to zero between lgr and lm (-0.05 ± 0.04). These results indicate that selection for increased lgr will lead to faster growing Piétrains without changing lean meat percentage.

Keywords: Genetic parameters, Reml, Lean growth rate, Piétrain, Swine

Introduction

Selection on lean tissue growth is mostly performed by means of an economic selection index including traits like growth rate and backfat thickness. However, lean growth rate in itself is also a selection (albeit a biological) index, as it is the product of average daily gain and the percentage of lean meat (Fowler *et al.*, 1976). The aim of this study is to estimate genetic parameters for the component traits growth rate and lean meat percentage and their product trait lean growth rate for the Belgian herdbook Piétrain population.

Material and Methods

Data

Data were obtained from 7 progeny testing stations belonging to the National Pig Breeders Association of Belgium (BEVA). The performance data of 18,755 purebred Piétrain barrows and gilts tested between 1984 and 1998 were used in the analysis. Collection of lean meat data started only in 1992. Lean meat percentage (and therefore also lean growth rate) was recorded for 7,908 animals. Tested pigs originated from 260 different herds, which were all members of BEVA. The structure of the data is shown in table 1.

Piglets started the testing period at about 25 kilograms liveweight. They were tested in pens of 4 littermates (2 barrows and 2 gilts) up to 100 kilograms liveweight. Animals were fed *ad libitum* during the whole testing period. A first diet, fed from the start of the testing period until the pigs reached 50 kilograms liveweight contained 18.4 % crude protein and 2,260 kilocalories of net energy. The second diet, fed from 50 kilograms liveweight until the end of the testing period contained 17.1 % crude protein and 2,300 kilocalories of net energy. At the end of the test, total feed intake over the testing period was recorded and pigs were slaughtered. Carcasses were weighed at around 30 minutes after killing. At the same time, lean meat percentage (lm) was measured by the skgii-apparatus (Casteels, 1989). Liveweight at the end of the test was calculated as carcass weight x 1.235 Lean growth rate is calculated as lgr = adg x 0.81 x lm, where adg is the average daily gain during the testing period. Growth rate is expressed as days on test. All traits (lm, lgr and days) are precorrected for sex differences and are adjusted to an initial liveweight of 25 kilograms and a final liveweight of 100 kilograms using correction factors as estimated by Geysen & Vandepitte (1998). After adding pedigree information the total number of animals (including pedigree information) in the analysis was 30,136.

Table 1.   Structure of the data set analysed.

	traits measured	
	days	lgr, lm[1]
pigs tested	18,755	7,908
barrows tested	8,826	3,580
gilts tested	9,929	4,328
no of sires	1,766	804
no of dams	4,537	1,989
no of litters	5,069	2,213
no of pigs/litter	3.7	3.6
no of sys[2]	285	136
no of pigs/sys	65.8	58.1

[1] lean growth rate, lean meat percentage
[2] station-year-season

Model

Genetic effects, common environmental litter effects, station-year-season effects and residual variance and covariance components were estimated using a reml algorithm on the basis of analytical gradients applied to a multiple trait animal model (Neumaier & Groeneveld, 1998). The analysis was performed using the VCE4 program (version 2.5). The model for each trait was :

$$y_{ijk} = \mu + h_i + c_j + a_k + e_{ijk}$$

where y_{ijk} = observation of animal k from litter j tested in herd-year-season i; μ = over-all mean; h_i = random effect of station-year-season i; c_j = random common environmental effect of litter j, a_k = random additive genetic effect of pig k; e_{ijk} = random residual effect associated to the ijk[th] observation. Phenotypic correlations between traits were calculated as Pearson correlation coefficients before adjusting the data for initial and final weight and precorrection for sex differences.

Results

Means, phenotypic variances (before and after precorrections) and variance components for the 3 traits studied are shown in table 2. Heritabilities and genetic and phenotypic correlations are given in table 3.

Table 2.   Means, phenotypic variances and variance components.

	lm (%)	lgr (g/day)	days (days)
mean[1]	63.3	299	133
variance			
total[1]	6.66	2,450	581
model[2]	5.17	2,195	444
residual	3.04	917	194
litter	0.08	185	54
sys	0.33	375	76
additive genetic	1.73	547	122

[1] before precorrections for initial and final weight and sex differences
[2] after precorrections for initial and final weight and sex differences

Table 3. Heritabilities (diagonal), phenotypic[1] (under diagonal) and genetic (above diagonal) correlations.

	lm	lgr	days
lm	0.33 ± 0.02	-0.05 ± 0.04	0.30 ± 0.03
lgr	-0.10	0.27 ± 0.02	-0.94± 0.01
days	0.31	-0.95	0.27 ± 0.01

[1] Pearson correlation coefficients, before precorrections for initial and final weight and sex differences. All phenotypic correlations are significantly different from zero at the 99 % probability level.

Discussion

Recent estimates of genetic parameters in the Piétrain breed are scarce, with the exception of Bidanel *et al.* (1994) and Bidanel & Ducos (1995). Heritability for growth rate is slightly lower than heritabilities found by the French authors. Heritability of lean meat percentage is much lower due to much lower phenotypic and genetic variance. Genetic and phenotypic correlations between growth rate and lean meat percentage are slightly more unfavourable in the present study.

Heritability of lean growth rate is equal to heritability of days of fattening, while genetic and phenotypic correlations between these traits are close to minus unit. Genetic and phenotypic correlation between lean growth rate and lean meat percentage are nearly zero. Very similar results were obtained by Mrode & Kennedy (1993). So, responses to selection on the product trait lgr will be dominated by 'days', which is the component trait with the largest variance (see also review of Simm *et al.*, 1987). For this reason, Cameron (1994) designed a selection criterion for lgr (based on average daily gain and backfat thickness) in order to achieve equal responses to selection for both component traits in a selection experiment in a Large White herd. He found an overall heritability (over 3 selection groups) for this selection criterion of 0.38. In a similar selection experiment in a Landrace herd, Cameron & Curran (1994) found an overall heritability of 0.25.

Lm and days are unfavourably correlated, while the genetic correlation between lgr and lm is very close to zero. Recalling the genetic correlation between lgr and days, this means that selection on lgr will lead to a genetic improvement of days, without a decrease in lm. These results also imply that no further genetic improvement for lm will be achieved if selection is based solely on lgr.

Acknowledgements

This work was financially supported by the Ministry of Small Enterprises, Traders and Agriculture (Brussels). A. D'Hooghe and F. Steyaert (both BEVA, Scheldewindeke) are gratefully acknowledged for providing the data.

References

Bidanel, J.P., A. Ducos, F. Labroue, R. Guéblez & C. Gasnier, 1994. Genetic parameters of backfat thickness, age at 100 kg and meat quality traits in Pietrain pigs. Ann. Zootech. 43 (2): 151-150

Bidanel, J.P. & A. Ducos, 1995. Variabilité et évolution génétique des caractères mesurés dans les stations publiques de contrôle de performances chez les porcs de race Piétrain. Journées Rech. Porcine en France, 27: 149-154

Cameron, N.D., 1994. Selection for components of efficient lean growth rate in pigs. 1. Selection pressure applied and direct responses in a Large White herd. Anim. Prod. 59: 251-262

Cameron, N.D. & M.K. Curran, 1994. Selection for components of efficient lean growth rate in pigs. 2. Selection pressure applied and direct responses in a Landrace herd. Anim. Prod. 59: 263-269

Casteels, M., 1989. Objectieve karkasbeoordeling binnen de Europese Gemeenschap. Verhandelingen van de Faculteit Landbouwwetenschappen te Gent 27: 38-55

Fowler, V.R., M. Bichard & A. Pease, 1976. Objectives in pig breeding. Anim. Prod. 23: 365-387

Geysen, D. & W. Vandepitte, 1998. Trimestriëel rapport 01/01/1999 - 31/03/1999. Research project no. S - 5832 section 3.

Mrode, R.A. & B.W. Kennedy, 1993. Genetic variation in measures of food efficiency in pigs and their genetic relationship with growth rate and backfat. Anim. Prod. 56: 225-232

Neumaier, A. & E. Groeneveld, 1998. Restricted maximum likelihood estimation of covariances in sparse linear models. Gen., Sel., Evol. 30: 3-26

Simm, G., C. Smith & R. Thompson, 1987. The use of product traits such as lean growth rate as selection criteria in animal breeding. Anim. Prod. 45: 307-316

Genetic parameters for colour traits and pH and correlations to production traits

S. Andersen & B. Pedersen

National Committee for Pig Breeding, Health and Production, Axeltorv 3, DK- 1609 Copenhagen V, Denmark

Summary

Meat colour in *m. longissimus dorsi* was measured as Minolta L, a and b values and visually assessed on the Japanese scale (Jap). In total 4,902 boars from the Landrace, Yorkshire and Duroc breeds, free from the Hal^n and the RN^- alleles, were measured. Ultimate pH, average daily gain (DG), lean meat percentage (LM) and feed efficiency (FE, individual feed intake recordings) were also measured. Estimated parameters differed only little between breeds. Heritabilities for Jap, L, b and pH were moderate (0.15 - 0.29), the heritability of a was high (0.52 - 0.57). Genetic correlations between production traits and Jap were very low ($|r_g| < 0.1$). Selection for darker meat was not found to be antagonistic to selection for production traits.

Keywords: Meat colour, pH, Heritability, Correlation, Production traits

Introduction

There is a consumer demand for darker meat and the aim of this study was to evaluate the genetic parameters for meat colour and the relationship between meat colour and the production traits in the Danish Landrace, Yorkshire and Duroc breeds.

Material and methods

Meat colour was measured on all slaughtered Landrace (LL), Yorkshire (YY) and Duroc (DD) boars which were tested at the test station Bøgildgård during a three year period. The boars were penned, 12-14 in each pen, fed *ad libitum* and feed intake was individually recorded using the ACEMA feed stations. All boars were slaughtered in the same plant. Transportation time was two hours and the boars rested one hour before slaughter. Meat colour was measured as lightness (L), redness (a) and yellowness (b) with the Minolta CR300 equipment approximately 24 hours after slaughter. Measurements were taken on a loin chop cut near the hip after a blooming period of 1 hour. Four measurements were taken on each chop and then averaged. Meat colour was also visually assessed using the Japanese scale (Nakai *et al.*, 1975) and ultimate pH was measured with pH x K21 from NWK-Binär in *Longissimus dorsi* approximately 48 hours after slaughter. Average daily gain (DG) from 30-100 kilograms was recorded as part of the performance test, which also included records of feed efficiency (kg feed / kg gain (FE)) in the test period and lean meat percentage (LM). Performance tested boars selected for AI and therefore not slaughtered at 100 kilograms obtained a lean meat percentage record from ultrasonic measurements of backfat. To take account of the selection effect it was decided that all boars, including non-slaughtered boars, should enter in the analysis of genetic correlations between colour and production traits. The halothane gene has been eradicated from the three breeds.

Genetic parameters for colour traits and pH were estimated for each breed in a multi trait animal model, using the VCE4 program (Neumaier & Groeneveld, 1998). The model included test batch (69, 70, 70 levels in LL, YY, DD), random effect of day of slaughter (92 levels) and a regression on carcass weight. The regression term was not included for pH. The pedigree was traced two generations backwards. The results are shown in tables 1 and 2. The heritabilities are calculated as additive genetic variance divided by the sum of additive genetic variance, residual variance and day-of-slaughter variance.

Genetic correlations to the production traits were estimated from two multi trait models for each breed, model 1 containing the traits DG, LM, FE, Jap and a; model 2 containing the traits DG, LM, Fe and L. Models for DG and LM included an effect of pen. Results are shown in table 3.

Discussion

The estimated genetic and phenotypic parameters for colour traits and pH differed only little between the three breeds Landrace, Yorkshire and Duroc. The heritability of visual colour, L and b were moderate 0.15 - 0.29, the heritability of the redness factor a was high, 0.52 -0.57.

These heritabilities are in agreement with most other studies. Van der Voort & Gibson (1997) found a heritability of 0.21 for visual colour, Dietl et al. (1993) report a heritability of 0.23. De Vries et al. (1994) report a heritability of L equal to 0.21 (SE = 0.05), Larzul et al. (1997) in a smaller study find the value 0.23 (SE = 0.08). Knapp et al. (1996) find breed specific values of colour measured by the Göfo equipment, $h^2 = 0.26$ in Yorkshire, $h^2 = 0.12$ in Landrace and $h^2 = 0.22$ in Piétrain. In an analysis of data from a selection experiment with Large White Sonesson et al. (1998) find considerably higher heritabilities: visual colour $h^2 = 0.72$ (SE = 0.08), Hunter L $h^2 = 0.73$ (SE = 0.08), Hunter a $h^2 = 0.52$ (SE = 0.07), and Hunter b $h^2 = 0.54$ (SE = 0.08).

The genetic correlations in table 2 show a high positive correlation between a and visual colour and high negative correlation between L and visual colour, whereas L and a are much less correlated which is also the case for b and Jap score. Note also that the a score is much less influenced by slaughter day than the other traits. Sonesson et al. (1998) find a somewhat different pattern of genetic correlations. They find a strong correlation, -0.70, between visual colour and b, and also a strong correlation, -0.67, between L and a. On the other hand they find a very weak correlation, -0.08, between a and b.

The genetic correlations r_g between visual colour and production traits show very low values $|r_g| < 0.1$. The genetic correlations between Minolta scores and production traits were low and inconsistent across breeds. De Vries et al. (1994) report genetic correlations between L and lean meat (-0.16), L and growth rate (0.21) and L and feed intake (-0.24) and they conclude that selection for high feed efficiency might reduce DFD incidence i.e. result in lighter meat. The general conclusion from the present study is selection for colour is possible and that selection for production traits will have little impact on colour traits.

Table 1. *Genetic variances (σA2), genetic correlations (above diagonal), residual correlations (below diagonal) of visual colour score (Jap), Minolta L, a and b scores and pH..*

Standard errors of genetic correlations ranging from 0.03 to 0.14
Standard errors of residual correlations ranging from 0.02 to 0.06
Standard errors of heritabilities ranging from 0.02 to 0.04

Landrace	Jap	L	a	b	pH	σ_A^2
Jap	**0.29**	-0.91	0.55	-0.20	0.28	0.22
L	-0.71	**0.22**	-0.18	0.50	-0.47	1.95
a	-0.14	0.49	**0.57**	0.68	-0.36	0.59
b	-0.53	0.82	0.65	**0.19**	-0.55	0.23
pH	0.58	-0.61	-0.14	-0.58	**0.17**	0.002

Yorkshire	Jap	L	a	b	pH	σ_A^2
Jap	**0.23**	-0.85	0.68	-0.03	0.36	0.19
L	-0.73	**0.15**	-0.28	0.50	-0.75	1.40
a	-0.11	0.44	**0.52**	0.66	-0.06	0.48
b	-0.49	0.77	0.70	**0.23**	-0.64	0.22
pH	0.46	-0.52	-0.36	-0.54	**0.16**	0.002

Duroc	Jap	L	a	b	pH	σ_A^2
Jap	**0.17**	-0.64	0.56	0.06	0.13	0.10
L	-0.67	**0.15**	0.12	0.59	-0.28	0.98
a	0.00	0.35	**0.54**	0.79	-0.38	0.53
b	-0.48	0.74	0.61	**0.27**	-0.57	0.28
pH	0.33	-0.50	-0.24	-0.48	**0.19**	0.003

Table 2. *Proportion of phenotypic variance explained by day of slaughter.*

Breed	Jap	L	a	b	pH
Landrace	0.10	0.12	0.02	0.18	0.17
Yorkshire	0.20	0.17	0.06	0.16	0.16
Duroc	0.19	0.13	0.01	0.13	0.20

Table 3. *Genetic correlations between production traits and colour traits. Standard errors of correlations ranging from 0.02 to 0.13.*

Prod. Trait	Colour trait	Breed		
		Landrace	Yorkshire	Duroc
DG	Jap	-0.04	0.08	0.07
LM	Jap	0.00	-0.04	0.07
FE	Jap	-0.03	-0.08	-0.04
DG	a	-0.10	-0.30	0.06
LM	a	-0.03	0.15	-0.07
FE	a	-0.04	-0.09	0.16
DG	L	0.12	-0.21	0.15
LM	L	-0.16	0.21	-0.26
FE	L	-0.02	-0.06	0.08

References

Dietl, G., E. Groeneveld & I. Fiedler, 1993. Genetic Parameters of Muscle Structure Traits in Pig. EAAP, Århus

Knapp, P., A. William, J. Sölkner, 1996. Genetic Parameters for Lean Meat Content and Meat Quality Traits in Different Types of Pig Breeds. EAAP, Lillehammer

Larzul, C., L. Lefaucheur, P. Ecolan, J. Gogué, A. Talmant, P. Sellier, P. Le Rog & G.

Mouin, 1997. Phenotypic and Genetic Parameters for Longissimus Muscle Fiber Characteristics in Relation to Growth, Carcass and Meat Quality Traits in Large White Pigs. J. Anim. Sci. 75: 3126-3137

National Pork Producers Council, 1995. Terminal Line Program Results. Genetic Evaluation

Neumaier, A. & E. Groeneveld, 1998. Restricted Maximum Likelihood Estimation of Covariances in Sparse Linear Models. Genet. Sel. Evol., 30: 3-26

Oksbjerg, N., J.S. Petersen, P. Henckel & S. Stoier, 1997. Meat Colour and Muscle Pigment in Danish Landrace Anno 1997 and Anno 1995. EAAP, Wien

Sonesson, A.K., K.H. de Greff, T.H.E. Meuwissen, 1998. Genetic parameters and trends of meat quality, carcass composition and performance traits in two selected lines of Large White Pigs. Livest.Prod.Sci. 57: 23-32

Voort van der, G. & J.P. Gibson, 1996. Estimation of Variance Components from the OPCAP DATA. Ontario Pork Carcass Appraisal Project Symposium

Vries de, A.G., P.G. van der Wal, G. Eikelenboom & J.W.M. Merks, 1994. Genetic Variance of Pork Quality in Yorkshire Populations. Livest.Prod.Sci. 40: 277-291

Gene effects on pork quality

Halothane gene effect on carcass and meat quality by use of Duroc x Piétrain Boars

H. Busk[*1], *A. Karlsson*[1] *& S.H. Hertel*[2]

Danish Institute of Agricultural Sciences, Dept. of Animal Product Quality, Research Centre Foulum, P.O.Box 50, DK-8830 Tjele, Denmark. [2]*PIC Denmark A/S, P.O.Box 7021, DK-9200 Aalborg SV, Denmark*

Summary

The halothane gene in pigs has in the heterozygote state (HAL-Nn) a positive effect on production and carcass traits. With regard to meat quality there are still problems with the water-holding capacity of meat in some breeds and breed-crosses. Duroc pigs and crosses are known to have a good meat quality compared to other breeds. Therefore, the aim of this project is to combine production traits with improved meat and carcass quality using halothane heterozygote cross-breed boars of the Piétrain and Duroc breeds (PxD). PxD boars (PIC 409) were mated to LxY sows. The offspring selected for the trial were equally distributed with Halothane negative homozygotes (HAL-NN) and heterozygotes (HAL-Nn). In total, 196 pigs were slaughtered at a high stress level. The results showed no differences in daily gain between the genotypes. With regard to carcass quality, the area of *m. longissimus dorsi* was 2.0 cm^2 less, the sidefat thickness 1.9 mm thicker and percentage of lean meat in the carcass 1.9 percent less for HAL-NN than for HAL-Nn. For meat quality, the pH values at 0 minutes, 45 minutes and 24 hours were higher, and the drip loss 3.7 percent units lower for HAL-NN than for HAL-Nn, but there were no differences in colour. In summary, the results showed no effect of the HAL- gene on daily gain, but a positive effect on carcass quality and a negative effect on the meat quality expressed as water-holding capacity.

Keywords: Carcass quality, Duroc, Halothane, Meat quality, Piétrain

Introduction

The main goals in pig breeding have for many years been to improve growth rate, feed conversion and carcass composition. There have been fewer efforts to improve meat quality parameters (water-holding capacity, colour, pH etc.) but the main contribution has been a reduction of PSE meat (Pale, Soft and Exudative). Improvements have been made, not only by selecting for better genotype, but also by providing better environments and feeding regimes. However, in the future, consumer demands and ongoing competition in the global market makes it necessary for the pig producers to include meat quality parameters in their breeding goals.

One way to improve the meat quality in pig production could be by using breeds with special characteristics in a crossing situation. It is well known that different breeds are different for production traits, carcass quality and meat quality. It is also well known that the halothane gene in a heterozygote state in pigs (HAL-Nn) has a positive effect on production and carcass traits. With regard to meat quality there are still problems with the water-holding capacity of meat in breeds and breed-crosses carrying the halothane gene (HAL-Nn; HAL-nn). Duroc pigs and crosses, where this breed is included, are known to have a positive effect on meat quality compared to other breeds. Therefore, the aim of this project was to investigate if it is possible to combine production traits with improved meat and carcass quality using halothane heterozygote cross-breed Piétrain and Duroc boars (PxD).

Material and methods

The experiment was carried out on pigs from a commercial farm using Piétrain x Duroc (PxD) boars (PIC 409) mated to Landrace x Yorkshire (LxY) sows. The offspring were tested for halothane status by use of the Hal-1843 DNA test, and then selected for the trial. In total, 196 pigs were included in the experiment, equally distributed for Halothane negative homozygotes (HAL-NN) and heterozygotes (HAL-Nn) and for sex (females and castrated males). The selected pigs were kept in pens with 12-18 pigs and group fed, and after 12 to 14 weeks they were sent to slaughter at the research abattoir at the Danish Institute of Agricultural Sciences.

To get a standardised stress level for the pigs at slaughter, they ran on a treadmill at a speed of 3.8 km/h for 10 minutes immediately prior to stunning using 80 % CO_2. After slaughter, the carcasses were stored at $4^{\circ}C$ in a chilling room. At 24 hours after slaughter, 100 grams of meat from *m. longissimus dorsi* (LD muscle) were cut in the region of the last rib for measuring driploss by the Honnikel bag method, where the sample is placed in a plastic bag for 48 hours at $4^{\circ}C$. The area of the LD muscle and sidefat thickness 8 cm off the midline at the last rib, were measured. The left side of the carcass was dissected for meat, fat and bone according to the Danish method. The pH was measured in the loin muscle at 1 and 45 minutes and at 24 hours after exsanguination. Meat colour was measured 24 hours after exsanguination as lightness (L*), redness (a*) and yellowness (b*) using a Minolta Chroma Meters CR-300. Pigment content (myoglobin) was measured as described by Oksbjerg *et al.* (1997).

The statistical calculations were carried out using the GLM procedure from SAS (1991). For the production traits, the fixed effects of HAL-genotype, sex, litter and pen were used in the model and the results were corrected to the same liveweight. For carcass and meat quality, the fixed effects of HAL-genotype, sex, boar and day of slaughter was used in the model and the results corrected to equal carcass weight. No interactions were seen.

Results and discussion

Because the pigs had not been fed individually, it was not possible to calculate the feed conversion. Daily gain has been calculated from 30 to 100 kilograms and LS-Means was 776 grams, and there was no difference between the HAL-genotypes. This is in agreement with results from other experiments (Larzul *et al.*, 1997; Lech *et al.*, 1996; Webb *et al.*, 1994).

The cold carcass weight was 81.2 kilograms for both HAL-genotypes. The organ's weights were registered just after slaughter and the results are shown in Table 1. The weight for heart, kidney and liver was found to be higher for the HAL-NN than for the HAL-Nn pigs, while there was no significant difference for the spleen.

Table 1.   LS-Means for HAL-NN, difference between HAL-genotypes and SED for organ
weights (gram).

| | HAL-genotype | | Standard Error | |
	NN	Nn-NN	of Difference	P
Heart	312	-10	3	*
Liver	1663	-55	15	*
Kidney	294	-12	4	*
Spleen	151	4	3	NS

*) P<0.05 NS) P>0.05

The most important results regarding carcass quality are shown in Table 2. The dressing percentage was equal for the two genotypes. In other experiments it has been shown that the HAL-Nn has a higher dressing percentage than the HAL-NN (Larzul et al., 1997; Gibson et al., 1996; Leach et al., 1996; McPhee & Trout, 1995) which could be explained by the lower organ weights for the HAL-Nn (Jones et al., 1988). Though the dressing percentage was equal for the genotype in this experiment, the total weight of the organs shown in Table 1 was 2420 grams for the HAL-NN and 2347 grams for the HAL-Nn, and in agreement with results from the mentioned experiments.

The area of the LD muscle was 2.0 cm^2 less, and the sidefat thickness 1.9 mm thicker for the HAL-NN than for the HAL-Nn. Results from the dissection showed 1.9 percent less lean meat, 1.3 percent more subcutaneous fat and 0.5 percent more intermuscular fat for the HAL-NN than for the HAL-Nn. There was no significant difference for skin and bone content between the two HAL-genotypes. The difference in lean meat percentage between the two genetic groups of pigs was at the same level and in agreement with other experiments (Leach et al., 1996; Gibson et al., 1996).

Table 2.   LS-Means for HAL-NN, difference between HAL-genotypes and SED for carcass
quality traits.

| | HAL-genotype | | Standard Error | |
	NN	Nn-NN	of Difference	P
Dressing %	79.9	0.4	0.24	NS
Area of L.D., cm^2	43.5	2.0	0.58	***
Sidefat thickn., mm	13.4	-1.9	0.40	***
% of the carcass				
Lean meat	59.2	1.9	0.41	***
Subcutaneous fat	18.0	-1.3	0.35	***
Intermuscular fat	6.3	-0.5	0.15	***
Skin	3.3	0.0	0.05	NS
Bone	13.2	-0.1	0.12	NS

***) P<0.001   NS) P>0.05

Concerning meat quality, the pH values at 1 minute, 45 minutes and 24 hours after slaughter were higher for the HAL-NN than for the HAL-Nn, which can be seen in Table 3. The big difference in pH$_{45}$ has also been seen in other experiments (De Smet et al., 1996).

The HAL-NN had less driploss than the HAL-Nn (3.7 percent points), which is at the same level as in other experiments (Leach et al., 1996: Sutton et al.; 1996 Lundstrom et al., 1989). There were no differences between HAL-genotypes in meat colour measured as lightness (L*), redness (a*) or yellowness (b*), and the pigment content did not differ between the two HAL-genotypes, which was confirmed by the a* value.

Table 3.   LS-Means for HAL-NN, difference between HAL-genotypes and SED for meat quality traits.

| | HAL-genotype | | Standard Error | |
	NN	Nn-NN	of Difference	P
Driploss, %	6.1	3.7	0.433	***
pH$_1$	6.61	-0.14	0.041	***
pH$_{45}$	6.39	-0.34	0.042	***
pH$_{24}$	5.61	-0.05	0.013	***
Lightness (L*)	50.6	0.5	0.380	NS
Redness (a*)	7.25	0.20	0.194	NS
Yellowness (b*)	5.49	0.19	0.155	NS
Pigment mg/g	0.756	0.025	0.024	NS

***) $P<0.001$ NS) $P>0.05$

Conclusion

In summary, the results showed no effect of the HAL-gene on daily gain, but a positive effect on carcass quality and a negative effect on the meat quality expressed as water-holding capacity and pH$_{24}$. The results are in general in agreement with results from other experiments. Compared to results in the literature, our data suggest that including 25 percent of the Duroc breed in slaughter pigs to improve the meat quality, does not, however, abolish the negative effect of the halothane gene on meat quality.

References

De Smet, S.M., H. Pauwels, S. De Bie, D.I. Demeyer, J. Callewier & W. Eeckhout, 1996. Effect of Halothane Genotype, Breed, Feed Withdrawal and Lairage on Pork Quality of Belgian Slaughter Pigs. Journal of Animal Science 74: 1854-1863

Gibson, J.P., R.O. Ball, B.E. Uttaro & P.J. O'Brien, 1996. The Effects of PSS Genotype on Growth and Carcass Characteristics. Proceedings of the Ontario Pork Carcass Appraisal Project Symposium, 35-38

Jones, S.D.M., A.C. Murray, A.P. Sather & W.M. Robertson, 1988. Body proportions and carcass composition of pigs with known genotypes for stress susceptibility fasted for different periods of time prior to slaughter. Canadien Journal Animal Science 68:139

Larzul, C., P. Le Roy, R. Guéblez, A. Talmant, J. Gogué, P. Sellier & G. Monin, 1997. Effect of the halothane genotype (NN, Nn, nn) on growth, carcass and meat quality traits of pigs slaughtered at 95 kg or 125 kg live weight. J. Anim. Breed. Genet. 114: 309-320

Leach, L.M., M. Ellis, D.S. Sutton, F.K. McKeith & E.R. Wilson, 1996. The Growth Performance, Carcass Characteristics, and Meat Quality of Halothane Carrier and Negative Pigs. Jounal of Animal Science 75: 934

Lundstrom, K., B. Essen-Gustovsson, M. Rundgren, I. Edfors-Lilja & G. Malmfors, 1989. Effect of Halothane Genotype on Muscle Metabolism at Slaughter and Its Relationship with Meat Quality: A Within-litter Comparison. Meat Science, 25: 251-263

McPhee, C.P. & G.R. Trout, 1995. The effects of selection for lean growth and the halothane allele on carcass and meat quality of pigs transported long and short distances to slaughter. Livestock Production Science 42: 55-62

Oksbjerg, N., J.S. Petersen, P. Henckel & S. Støier, 1997. Meat colour and muscle pigment in Danish Landrace Anno 1976 and Anno 1995. 48th EAAP, Wien. Austria. pp4

SAS. 1991. SAS User`s Guide: Statistics (Version 6, 4th Ed.) SAS Inst. Inc., Cary, NC

Sutton, D.S., M. Ellis, Y. Lan, F.K. McKeith & E.R. Wilson, 1997. Influence of Slaughter Weight and Stress Gene Genotype on the Water-holding Capacity and Protein Gel Characteristics of Three Porcine Muscles. Meat Science 46: 173-180

Webb, A.J., B. Grundy & P. Kitchin, 1994. Within-Litter Effect of the Hal-1843 Heterozygote on lean growth in pigs. Proceedings of the 5th World Congress on Genetics Applied to Livestock Produce, 17: 421

In vivo and post mortem changes of muscle phosphorus metabolites in pigs of different Malignant Hyperthermia Genotype

M. Henning[1], U. Baulain[1], G. Kohn[1] & R. Lahucky[2]

[1]*Institut für Tierzucht und Tierverhalten (FAL), D-31535 Neustadt-Mariensee, Germany*
[2]*Research Institute of Animal Production, 94992 Nitra, Slovakia*

Summary

Stress susceptibility in swine causes a significant economic loss due to preslaughter death or poor meat quality. Different MHS (Malignant Hyperthermia Syndrome) genotypes - Piétrain (nn), heterozygous German Landrace (Nn) and homozygous stress resistant German Landrace (NN) were investigated by means of non invasive ^{31}P-spectroscopy. Changes of inorganic phosphate (P_i), phosphocreatine (PCr) and ATP in *M. biceps femoris* were analysed during halothane exposure. Intracellular pH and magnesium (Mg^{++}) which is a calcium antagonist and activator for ATP were estimated from the spectra. As expected nn genotypes show dramatic alterations in their muscle metabolism after administration of halothane. The positive reaction (muscle contraction and hyperthermia) often requires interruption of the experiment. A significant decrease of PCr and a corresponding increase of P_i were observed. Intracellular pH declined from 7.1 to 6.4, Mg^{++} increased by more than 100%. Nn and NN pigs did not react to halothane administration. At rest intracellular Mg^{++} was significantly higher in nn pigs as compared to NN and nN. In heterozygous pig's muscles intermediate concentrations were found, higher than in NN pigs but still significantly lower than in nn genotypes. *Post mortem* studies however yielded significant differences between nN and NN carcasses in time course changes of PCr and P_i as well as the rate of pH decline.

Keywords: Pigs, Stress, Meat quality, MHS gene, MR spectroscopy, nN

Introduction

Stress susceptibility in swine is caused by the MHS (Malignant Hyperthermia Syndrome)-gene which induces disorders in muscle metabolism. A mutation in the "foot region" of the ryanodine receptor reduces its sensitivity to magnesium (Fujii *et al.*, 1991). Therefore a high intracellular Ca^{++} concentration does not inactivate the ryanodine receptor in order to avoid uncontrolled muscle contractions. These contractions are consuming ATP which can be regenerated by phosphocreatine, aerobic glycolyses and, if stress is continued, by anaerobic glycolyses. This accumulates lactate and leads therefore to a pH decline with, in severe cases, denaturation of plasma and membrane proteins (Bickhardt *et al.*, 1972, Fiedler *et al.*, 1993).

This combination of conditions results in an exaggeration of the muscle to meat transformations that normally occur and the muscles usually become pale in colour, soft in texture and exudative (PSE) with lowered processing yields, increased cooking losses and reduced juiciness (Judge *et al.*, 1989).

Approximately 80% of the slaughter pigs in Germany are crossbreds. Most of them are of heterozygous MHS genotypes in order to produce a lean carcass with a stress resistant pig. But still the heterozygous animals produce about 10 % PSE carcasses (Glodek, 1999).

135

Therefore this group of animals was the most interesting one in the study presented on muscle metabolism.

Material and Methods

For the investigation German Landrace (nN=16, NN=12) and Piétrain pigs (nn=7) were available. Genotype was determined by MHS gene test according to Fujii *et al.*, 1991. The tests were carried out at the Institute of Animal Breeding and Genetics in Göttingen.

[31]P-spectroscopy was done in a 1.5 T whole body tomograph (BRUKER Germany) in Mariensee. Phosphorus spectra were obtained at a resonance frequency of 25.8 MHz with a double tuned surface coil (10 cm in diameter).

Pigs were sedated with an i.m. injection of Tilest 500 (PARKE-DAVIES) before they were moved into the tomograph. The surface coil was placed on the left hind leg to receive signals from *M. biceps femoris*.

Prior to halothane administration five spectra were recorded. The mean value (signal intensity) of PCr served as reference for consecutive spectra. Relative concentrations of metabolites were determined by integrating the areas of the corresponding signals (Fig. 1).

Intracellular pH was calculated according to the Henderson-Hasselbach-equation regarding the chemical shift δ of P_i to PCr (in ppm). For intracellular Mg++ the distance of resonance signals for α- and β-ATP are read, pH correction is essential. Further details of evaluation are described by Kohn (1997).

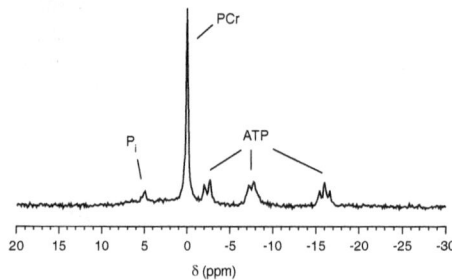

Figure 1. ^{31}P spectrum of the M. biceps femoris of a pig at rest. The peaks from left to right are: inorganic phosphate (Pi), phosphcreatine (PCr), γ-, α- and β-ATP.

Results

Figure 2 shows a typical reaction of relative metabolite concentrations in an nn-pig which recovered after a positive halothane reaction. The onset of this reaction is seen as a decrease of the PCr signal and an increase of P_i. This occurred at t = 15 min, i.e. 10 minutes after exposure to 2% halothane. PCr content dropped down to 36% of its initial value within nine minutes. Simultaneously concentration of P_i raised from 5.4% to 61.0 %. It is clearly visible that at t = 25 minutes the relative concentration of PCr increases again and P_i declines which indicates the recovery of the animal.

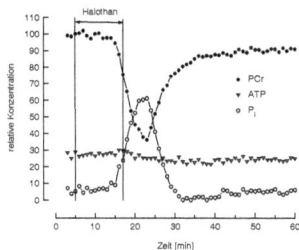

Figure 2. Changes of phosphorus metabolite concentration in the M. biceps femoris of a MHS positive pig (nn) which survived halothane exposure.

The relation of phosphocreatine to inorganic phosphate (PCr/P_i) is a measure of the cell's energy supply. Figure 3 shows the alteration of this quotient under halothane exposure. Genotype differences are clearly distinct. PCr/P_i-values at rest from Nn-genotypes were significantly higher (8.56 ± 2.32 units) than in homozygous negative pigs (7.17 ± 1.62 units), but still lower than values of nn-genotypes (12.36 ± 0.71 units).

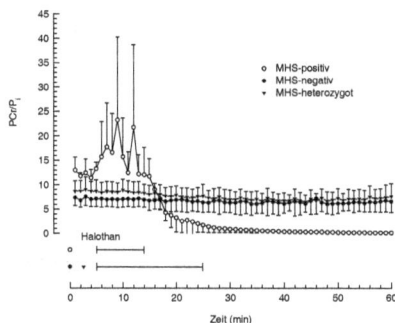

Figure 3. Changes of PCr/P_i in the M. biceps femoris of pigs of different MHS genotype under halothane exposure.

The alterations of intramuscular pH p.m. in carcasses of NN- und Nn-pigs are presented in Figure 4. During the first hour after slaughter it decreased by $3.6 \cdot 10^{-3}$ in NN, and by $8.5 \cdot 10^{-3}$ units/min. in nN genotypes. In the second hour the rate was $3.0 \cdot 10^{-3}$ (NN) and $5.3 \cdot 10^{-3}$ unit/min. in Nn, respectively.

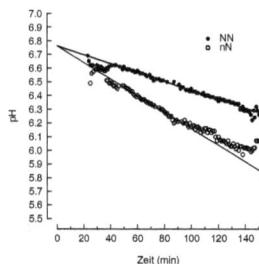

Figure 4. Post mortem changes of pH in M. biceps femoris of pigs of different genotypes.

Mg^{++} concentrations are listed in Table 1. Significant differences are already seen at rest between NN and nn genotypes and also between nn and nN pigs. The 10% higher concentration of nN compared to NN genotypes shows a tendency to metabolic problems, but it is not significant.

Table 1. Intracellular Mg^{++} concentration in M. biceps femoris of pigs in µmol/l (Means and SD of five x n/genotype spectra).

genotype	Mg^{++} concentration (µmol/l)		
NN (n=12)	387.0 [a]	±	83.1
nN (n=16)	425.8 [b]	±	136.5
nn (n=7)	495.2 [ab]	±	158.9

[a,b] Values with identical letters are significantly different.

Conclusions

^{31}P-spectroscopy allows a non invasive online determination of phosphorus metabolites in muscle tissue of pigs.

Halothane exposure did not induce a visible reaction in energy metabolism in muscles of NN and nN genotypes. Distinct reactions such as a decline in PCr-concentration and a corresponding increase of inorganic phosphate (P_i) are only seen in pigs of nn-genotype.

Intracellular Mg^{++} is significantly higher in M. biceps femoris in nn pigs than in nN- and NN-genotypes, values of nN-pigs being intermediate.

Post mortem investigations however yielded significant differences between carcasses of NN- and nN-genotypes. Carcasses of heterozygous pigs developed an accelerated decline of PCr and thus a faster pH-decrease.

Results confirm that nN pigs are still potentially susceptible for development of PSE meat.

References

Bickhardt, K., H.J. Chevalier, W. Giese & H.J. Reinhardt, 1972. Akute Rückenmuskelnekrose und Belastungsmyopathie beim Schwein. Adv. Vet. Med. 18

Fiedler, I., K. Ender, M. Wicke, G. v. Lengerken, 1993. Zusammenhänge zwischen der Mikrostruktur des Muskelgewebes bei Schweinen der Landrasse und ihrer Stressempfindlichkeit. Arch. Tierz. 36: 525-538

Fujii, J., K. Otsu, F. Zorsato, S. de Leon, V.K. Khanna, J.E: Weiler, P.J. O'Brian & D.H. MacLennan, 1991. Identification of a mutation in porcine ryanodin receptor associated with malignant hyperthermia. Science 253: 448-451

Glodek, P., 1999. personal communication

Kohn, G., 1997. In-vivo-Untersuchungen zur Muskelphysiologie von Schweine mit Hilfe der Magnet-Resonanz-Spectroskopie. PhD Thesis in chemistry, University of Bremen, Germany

Judge, M.D., E.D. Aberle, J.C. Forrest, H.B. Hedrick & R.A. Merkel, 1989. Principles of meat science. Kendall/Hunt Publ. Company

Interactive effects of the HAL and RN major genes on carcass quality traits in pigs: preliminary results

P. Le Roy[1], C. Moreno[1], J.M. Elsen[2], J.C. Caritez[3], Y. Billon[3], H. Lagant[1], A. Talmant[4], P. Vernin[4], Y. Amigues[5], P. Sellier[1] & G. Monin[4]

[1] *INRA, SGQA, 78352 Jouy en Josas cedex, France*
[2] *INRA, SAGA, BP 27, 31326 Castanet Tolosan cedex, France*
[3] *INRA, Domaine du Magneraud, 17700 Surgères, France*
[4] *INRA, SRV, Theix, 63122 Saint Genès Champanelle, France*
[5] *INRA, LABOGENA, 78352 Jouy en Josas cedex, France*

Summary

An experiment was set up in order to estimate the interactive effects of the two major genes HAL (N and n alleles) and RN (rn^+ and RN^- alleles) on carcass quality traits in pigs. Pigs from each of the nine combined RN-HAL genotypes were recorded for growth, body composition and meat quality traits. The results showed that major effects of HAL and RN are essentially additive for carcass composition traits but significant HAL by RN interaction effects occur for most of the meat quality traits. In the latter case, two situations were encountered, either a "snowball" interaction effect when the effect of one mutation, n or RN^-, is increased by the other one, or a "lessening" interaction effect when the effect of one mutation is decreased by the other one. For example, the adverse effect of the n allele on pH_1 (or drip loss) was greater in carriers than in non-carriers of RN^-, whereas the adverse effect of the RN^- allele on Napole technological yield (or cooking loss) was smaller in nn than in NN pigs.

Keywords: Pigs, Meat quality, RN gene, Halothane, Major gene

Introduction

The major effects of RN (rn^+ and RN^- alleles) and HAL (N and n alleles) genes on carcass quality traits in pigs have been repeatedly shown (for review see, Sellier, 1998). Both genes have pleiotropic effects on lean meat content in the carcass and on meat quality. Halothane reactors, i.e. homozygotes nn, typically produce heavier, shorter and leaner carcasses, but also PSE pork, as halothane sensitivity primarily induces an acceleration of the *post mortem* muscle pH fall. The n allele acts in an approximately additive way regarding lean content and with significant recessiveness for most meat quality traits, the Nn genotype being intermediate but closer to NN than to nn genotype. The RN^- carriers have a higher muscle glycogen content than non-carriers. Consequently, they show a larger extent of the *post mortem* fall in muscle pH and they give acid meat. The RN gene also seems to affect in an additive manner the ratio lean/fat in the carcass.

Knowing these major individual effects of RN and HAL on most carcass quality traits, it was interesting to estimate their interactive effects. An experimental design was set up with this objective, pigs of the nine combined RN-HAL genotypes being produced and recorded for a number of performance traits. This short communication reports preliminary results concerning some of the main traits pertaining to growth, body composition and meat quality.

A complete analysis, including an extensive protocol for a better assessment of technological and sensory quality traits, is in progress.

Material and methods

Experimental design

The experiment was carried out in 1996-1998 on the Le Magneraud INRA farm (Surgères, Charente Maritime, France). Animals were from the INRA protocol started in 1990 to study the RN gene (Le Roy et al., 1999) using pigs of the Laconie composite line. This line of the Pen ar Lan breeding company (Maxent, Ille et Vilaine, France) was originally founded with Hampshire, Piétrain and Large White blood in equal proportions. To obtain pigs of the nine combined RN-HAL genotypes, 4 males and 20 females (RN⁻RN⁻,Nn) and 4 males and 22 females (rn⁺rn⁺,Nn) were mated. The RN genotypes of these parents were deduced from a progeny test for Napole technological yield (Le Roy et al., 1994) in a preceeding step of the design. The HAL genotypes were determined using a DNA-test (Dalens & Runavot, 1993). Thus, in any litter, all piglets were of the same a priori known RN genotype and of the three HAL genotypes, in proportions 25% for NN and nn, and 50% for Nn, as determined by molecular genotyping.

From 844 pigs recorded for growth performance, 447 were slaughtered and standardised carcass cutting was carried out. From a part of these animals (expected to be 40 pigs per genotype), meat quality measurements were taken. The numbers of animals recorded in each RN-HAL genotype and for each group of measured traits are given in table 1.

Table 1.   Numbers of pigs recorded for each group of traits.

| Genotype RN | rn⁺rn⁺ | | | RN⁻rn⁺ | | | RN⁻RN⁻ | | |
HAL	NN	Nn	nn	NN	Nn	nn	NN	Nn	nn
Growth	86	148	47	77	133	41	85	175	52
Carcass	45	70	30	53	69	26	64	66	24
Meat quality	41	48	29	42	48	26	48	47	24

Traits

Average daily gain was recorded individually from 30 to 100 kilograms liveweight. Feed consumption was recorded on a pen basis to calculate the food conversion ratio from 30 to 100 kilograms liveweight. Pigs were slaughtered at 100 kilograms liveweight in a commercial abattoir. On the day after slaughter, the carcass was weighed to calculate killing out percentage. Carcass length and midline backfat thickness, at shoulder, back and rump levels, were measured. Then, the half carcass was divided into seven joints which were weighed (Anonymous, 1990) ham, loin, shoulder, belly, backfat, leaf fat and feet. Physico-chemical characteristics of muscles differing by their metabolic and contractile properties were recorded: pH_1 at 35 minutes *post mortem* on the *longissimus* (LD), *semimembranosus* (SM) and *semispinalis capitis* (SC) muscles; ultimate pH (pH_u), colour (L*-value) on the LD, SM, SC, *adductor femoris* (AD), *biceps femoris* (BF), *gluteus superficialis* (GS) and *gluteus profondus* (GP) muscles, and imbibition time on the LD and BF muscles at 24 hours *post mortem*. The Napole curing-cooking yield (RTN) was measured on a 100 gram sample of the SM muscle (Naveau et al., 1985). Drip loss and cooking loss were recorded on one slice of meat according to Honikel (1987).

Statistical methods

Analyses of variance were performed using the SAS GLM procedure, traits being analysed successively in univariate models. The RN genotype (3 levels), the HAL genotype (3 levels) and their interaction were included in the model as fixed effects. The effects of sex (2 levels, castrated male and female) and series of measurements, i.e. fattening batch for growth and carcass composition traits (12 levels) and date of slaughter for meat quality traits (30 levels), were also taken into account. Initial weight for average daily gain and liveweight at slaughter for carcass and meat quality traits were included as covariates.

Results

The effect of the RN genotype by HAL genotype interaction was not significant on average daily gain and food conversion ratio, the individual effects of the two genes not being significant on either of these growth traits. With regard to carcass quality, individual effects of both genes were confirmed: on killing out percentage for HAL; on backfat thickness at shoulder, back and rump levels for RN, only at rump level for HAL; on carcass length for RN and HAL; on weights of joints, except the head weight, for RN and HAL. However, none of the RN by HAL interaction effects was significant at the 5 % level.

Table 2.   Traits significantly influenced by the RN genotype by HAL genotype interaction at the 5% level: least squares means for the RN-HAL combined genotypes[1].

Genotype RN	rn^+rn^+			RN^-rn^+			RN^-RN^-		
HAL	NN	Nn	nn	NN	Nn	nn	NN	Nn	nn
	pH$_1$ 35 minutes *post mortem*								
LD[2]	6.92 [a]	6.67 [b]	6.22 [c]	6.83 [a]	6.63 [b]	6.09 [cd]	6.89 [a]	6.69 [b]	6.02 [d]
SM[2]	6.85 [a]	6.62 [b]	6.15 [d]	6.65 [b]	6.46 [c]	5.72 [e]	6.75 [ab]	6.51 [c]	5.76 [e]
SC[2]	6.66 [a]	6.69 [a]	6.62 [a]	6.67 [a]	6.64 [a]	6.29 [b]	6.75 [a]	6.64 [a]	6.45 [b]
	L* 24 hours *post mortem* (scale 0-100)								
LD	47.7 [a]	51.6 [bc]	54.8 [d]	51.0 [bc]	52.3 [b]	56.2 [d]	48.4 [a]	49.9 [c]	56.4 [d]
	Drip loss (%)								
LD	3.3 [a]	3.4 [a]	3.7 [ac]	4.1 [ac]	5.0 [bc]	5.5 [b]	4.1 [ac]	4.6 [c]	5.8 [b]
	Cooking loss (%)								
LD	27.4 [a]	29.1 [ac]	29.3 [ac]	31.1 [b]	30.8 [bc]	30.7 [bc]	32.3 [b]	31.4 [b]	31.1 [bc]

[1]   Standard errors of least squares means were between 0.03 and 0.04 for pH$_1$, 0.5 and 0.8 for L*-value, 0.2 and 0.3 for drip loss, 0.4 and 0.7 for cooking loss

[2]   Muscles: LD=*longissimus*; SM=*semimembranosus*; SC=*semispinalis capitis*

[abcd] Means with the same letter did not differ at the 5% level

On the other hand, significant HAL by RN interaction effects were observed on several meat quality traits. For these traits, least squares means for the RN-HAL combined genotype effect are given in table 2. The pH$_1$ of meat was very highly affected by the two genes. For pH$_u$, only the effect in the LD muscle remained significant for HAL, whereas measurements on the seven muscles were all very highly affected by the RN genotype. The interaction between the two genes was highly significant for the pH$_1$ of the LD, SM and SC muscles. The well known effect of HAL on the pH fall rate was increased in the RN$^-$ carriers. Moreover a RN effect on the pH$_1$ was revealed only in homozygotes nn (table 2). On the contrary, no interaction was observed in meat pH at 24 hours *post mortem*. The L*-value was also influenced by the two genes, except in GP and SC muscles for HAL and in BF muscle for RN. However, the RN by HAL interaction was significant only for the LD muscle, the difference between the homozygotes NN and nn being significantly lower in the heterozygotes RN⁻rn⁺ than in the two other RN genotypes.

Finally, the two mutations decreased significantly the water holding capacity of fresh meat as assessed by either imbibition time or drip loss. The interaction effect was significant for drip loss, the pattern being the same as that for pH_1, i.e. one mutation, n or RN⁻, increased the effect of the other one. Cooking loss was affected only by RN whereas the RTN was significantly influenced by both genes. However, the interaction effect was highly significant on cooking loss but significant only at the 10% level on RTN. For these two traits, the interaction effect was opposite to that for pH_1 or drip loss, the difference between RN⁻ carriers and non-carriers being smaller in nn than in NN pigs. Thus, in this case, the effect of one mutation was to decrease the effect of the other one.

Conclusion

This comparison of the nine combined RN-HAL genotypes confirmed our previous knowledge of the individual effects of these two major genes. Furthermore, if no interaction effect was found on body composition traits, there were significant interactive effects on several meat quality traits. Two situations were then encountered, either a "snowball" or a "lessening" interaction effect. Thus, these experimental observations allowed a better understanding of the PSE or acid meat syndromes, showing the respective and interactive roles played by stress susceptibility (HAL) and muscle glycogen concentration (RN) in the determination of pig meat quality.

References

Anonymous, 1990. Résultats du 16ème test d'évaluation des performances de croissance, de composition corporelle et de qualité de la viande des produits terminaux des schémas de sélection et croisement. Techni-porc, 13: 29-45

Dalens, M. & J.P. Runavot, 1993. Test moléculaire pour le dépistage du gène de la sensibilité à l'halothane chez le porc. Techni-porc, 16: 17-20

Honikel, K.O., 1987. How to measure the water-holding capacity of meat ? Recommendation of standardized methods. In: Evaluation and control of meat quality in pigs, Martinus Nijhoff, Dordrecht, The Netherlands, 129-142

Le Roy, P., J.C. Caritez, J.M. Elsen & P. Sellier, 1994. Pigmeat quality: experimental study on the RN major locus. In: Proceedings of the 5th World Congress on Genetics Applied to Livestock Production, University of Guelph, Canada, 19: 473-476

Le Roy, P., J.M. Elsen, J.C. Caritez, A. Talmant, H. Juin, P. Sellier & G. Monin, 1999. Comparison between the three RN genotypes for growth and carcass quality traits in pigs. Submitted to Genet. Sel. Evol.

Naveau, J., P. Pommeret & P. Lechaux, 1985. Proposition d'une méthode de mesure du rendement technologique: la méthode "Napole", Techni-porc, 8: 7-13

Sellier, P., 1998. Genetics of meat and carcass traits. In: The genetics of the pig, Rothschild, M.F. & A. Ruvinsky (editors), CAB International, Oxon, UK, 463-510

Meat quantity to meat quality relationships when the RYR1 gene effect is eliminated

J. Kortz [1], W. Kapelanski[2]*, S. Grajewska[2], J. Kuryl[3], M. Bocian[2] & A. Rybarczyk[1].

[1] Agricultural University, Dr Judyma 24, 71-460 Szczecin
[2] University of Technology and Agriculture, Mazowiecka 28, 85-084 Bydgoszcz
[3] Institute of Genetics and Animal Breeding, Polish Academy of Science, 05-551 Mrokow, Jastrzebiec, Poland

Summary

Meat quality traits and carcass characteristics of 33 NN, 27 Nn and 30 nn pigs were determined. Overall (r_o) and intragroup (r_i) correlations between the meat quality and quantity traits were computed. Comparison of the magnitude of r_o and r_i correlations enables the anticipation of the real RYR1 gene effect on meat quality and quantity. The results suggest that the negative relationship between meat quality and quantity depends upon the linked RYR1 gene effect on meat deposition and on meat quality.

Keywords: Pigs, Lean content, Meat quality, RYR1 genotype

Introduction

The development of deterioration in meat quality in stress susceptible pigs is generally accepted as the genetic effect of the malfunctioning of the calcium release channel in the muscle reticulum due to single point mutation in the RYR1 gene (Cheah *et al.*, 1994; Lahucky *et al.*, 1997; Sellier & Monin, 1994). Two of the three RYR1 genotypes, heterozygous Nn and especially homozygous nn, are associated with the MH syndrome and with greater carcass lean content, larger longissimus muscle area and thinner backfat (Rempel *et al.*, 1995; Zhang *et al.*, 1992). The association between meat quantity and quality is probably a consequence of linkage of some gene loci which control the quantity and quality of meat (Hardge *et al.*, 1997; Kuryl *et al.*, 1998).

The aim of the study was to prove if the significant relationship between meat quantity and quality will exist when considered in all experimental pigs (overall correlation) or will exist irrespective of pig genotype (intragroup correlation).

Material and methods

The animals and the method used were the same as in the former paper (Kapelanski *et al.*, 1999).

Results and discussion

Meat quality traits for the three RYR1 genotypes are given in Table 1. Some significant (P < 0.01) deterioration of meat colour and water holding capacity towards PSE was proved in nn pigs. *Post mortem* changes in muscle pH were the fastest in nn pigs when compared with NN and Nn (pH$_1$ 5.75 versus 6.65 and 6.47). Ultimate pH (pH$_u$) also was the lowest in nn pigs (5.37 versus 5.45 and 5.42). The meat quality index I$_2$ based on the proportion of pH$_u$ to pH$_1$ values used to discern the normal from PSE muscle (Kortz, 1986) was also inferior in meat of nn pigs (P < 0.01). The most faithful quality indicator used for meat classification is the Q value which is composed of eight independent variables such as pH$_1$, pH$_u$, colour parameters, water holding capacity indices and sensoric meat assessment (Grajewska et al., 1984). A higher numerical value of Q indicates higher meat quality.

The general assessment of meat quality based on the Q value indicated significant differences between RYR1 genotypes of pigs mainly between recessive homozygotes nn and NN and Nn pigs. This is consistent with data reported by numerous authors (Sellier & Monin, 1994; Fisher & Mellett, 1997).

Chemical components of meat, i.e. water, fat and protein, did not show significant differences between pig genotypes, with the exception of intramuscular fat content. Meat of nn pigs had a minimal fat content. The result is in agreement with data of other authors (Zhang et al., 1992; Wicke et al., 1998). The RYR1 gene effect on intramuscular fat content seems to be linked with another genes controlling the body fat deposition.

Table 1. Meat characteristics.

Traits	Genotype		
	NN	Nn	nn
Number, no	33	27	30
Meat pigments, pom	46.48 ± 2.05	52.92 ± 2.31	47.34 ± 2.17
Dom. wavelength, nm	584.5[a] ± 0.27	585.5[b] ± 0.30	584.4[a] ± 0.28
Colour saturation, %	20.8[A] ± 0.6	22.2[AB] ± 0.6	23.9[B] ± 0.6
Colour lightness, %	23.4[A] ± 0.6	22.4[A] ± 0.7	27.7[B] ± 0.6
WHC, free water, %	21.7[A] ± 0.5	21.5[A] ± 0.6	24.6[A] ± 0.6
Plasticity, cm^2	2.06[aA] ± 0.03	1.94[bA] ± 0.04	1.63[B] ± 0.04
pH$_1$	6.65[Aa] ± 0.06	6.47[bA] ± 0.07	5.75[cB] ± 0.06
pH$_u$	5.45[aA] ± 0.01	5.42[bA] ± 0.01	5.37[cB] ± 0.01
Q	3.02[A] ± 0.05	2.95[A] ± 0.06	2.39[B] ± 0.06
I$_2$	4.29[aA] ± 0.15	3.85[b] ± 0.17	2.45[cB] ± 0.16
Water, %	74.17 ± 0.11	73.91 ± 0.12	74.11 ± 0.11
Fat, %	1.10[a] ± 0.07	0.84[b] ± 0.08	0.77[ab] ± 0.07
Protein, %	22.92 ± 0.12	23.38 ± 0.14	23.16 ± 0.13

a, b, c - Significant at P < 0.05; A, B - P < 0.01

Carcass characteristics are given in Table 2. As in the case of meat traits, significant differences in backfat thickness, loin eye area and predicted lean content were stated between nn genotype and others (P < 0.01).

The overall and intragroup correlations are given in Table 3. Data reflects the relationship between meat quantity and quality traits. The diminishing magnitude of correlation coefficients, r$_i$ versus r$_o$ may be interpreted as a real RYR1 gene effect exerted on relationships between correlated traits. As can be seen, a negative relationship between meat quantity and particular meat quality traits was not always of the same range. Lighter meat colour and lower meat quality index Q were associated with more meaty pigs (r$_o$ significant at

P < 0.01) and depended on linked genotype effects. If the relationship was considered within RYR1 genotypes, i.e. when the RYR1 gene effect is eliminated, then intragroup correlation coefficients (r_i) lost their significance.

Table 2. Carcass characteristics.

Traits	Genotype		
	NN	Nn	nn
Backfat thickness, mm	$27.8^A \pm 1.1$	$28.7^A \pm 1.2$	$18.8^B \pm 1.2$
Loin eye area, cm^2	$38.6^A \pm 1.4$	$39.0^A \pm 1.6$	$55.3^B \pm 1.5$
Predicted lean content, % (UFOM)	$45.7^A \pm 0.9$	$44.8^A \pm 1.0$	$55.6^B \pm 1.0$

A, B - Significant at P < 0.01

Table 3. Overall (r_o) and intragroup (r_i) correlation coefficients between meat quality traits and carcass characteristics.

Correlated traits	Av. backfat thickness, mm		Loin eye area, cm^2		Predicted lean content, %	
	r_o	r_i	r_o	r_i	r_o	r_i
Meat pigments, ppm	-0.15	-0.15	0.06	0.02	-0.23	-0.21
Dom. wavelength, nm	0.13	0.02	-0.22	-0.16	**-0.27**	-0.21
Colour saturation, %	-0.25	0.07	0.24	-0.01	0.20	-0.03
Colour lightness, %	**-0.43**	-0.16	**0.50**	0.21	**0.50**	0.22
WHC, free water, %	-0.14	0.10	**0.30**	0.06	0.21	-0.07
Plasticity, cm^2	**0.47**	0.16	**-0.63**	**-0.31**	**-0.52**	-0.17
pH$_1$	**0.55**	0.26	**-0.64**	**-0.26**	**-0.63**	**-0.31**
pH$_u$	0.18	-0.08	-0.24	0.10	-0.20	0.15
Q	**0.46**	0.14	**-0.56**	**-0.19**	**-0.53**	-0.14
I$_2$	**0.51**	0.28	**-0.59**	**-0.29**	**-0.60**	**-0.36**

Coefficients in bold type are statistically significant at P < 0.01

A less distinct situation existed when pH$_1$ and I$_2$ values were examined. Acidity of muscle was significantly related to higher lean content and in a lesser degree to the RYR1 gene effect, since the magnitude of correlation coefficients r_i versus r_o was only diminished. In conclusion, it may be said that negative relationships between meat quality and high lean content is mediated by the RYR1 gene status of pigs and depends on the frequency of the nn genotype.

References

Cheah, A.M., K.S. Cheah, R. Lahucky, L. Kovac, H.L. Kramer & C.P. McPhee, 1994. Identification of halothane genotypes by calcium accumulation and their meat quality using live pigs. Meat Sci. 38: 375-384

Fisher, P. & F.D. Mellett, 1997. Halothane genotype and pork production. 1 Growth, carcass and meat quality characteristics. S. Afr. J.Anim. Sci. 61: 109-114

Grajewska, S., J. Kortz & J. Rozyczka, 1984. Estimation of the incidence of PSE and DFD in pork. Proc. Scient. Meeting Biophysical PSE-muscle analysis. Vienna, 72-89

Hardge, T., K. Kopke, G. Leuthold, K. Wimmers & Th. Paulke, 1997. Differences in allele frequencies of candidate genes for growth, carcass value and meat quality between extreme phenotypes of commercial pig breeds. 48th Annual Meeting of EAAP, August 25-28, Vienna

Kapelanski, W., J. Kortz, J. Kuryl, T. Karamucki & M. Bocian, 1999. Correlations between growth rate, slaughter yield and meat quality traits after the elimination of RYR1 gene effect. 50th Annual Meeting EAAP, Zurich, Switzerland, 147-150

Kortz, J., 1986. An attempt of appointment the synthetic index of pork meat quality as criterion for distinguish the normal, PSE and DFD muscle. Hab. Dissertation No 100. Ed. Agricultural Academy Szczecin

Kuryl, J., M. Zurkowski, M. Rozycki, M. Duniec, M. Pierzhala, A. Kossakowska, A. Janik, M. Kamyczek, M. Szydlowski, I. Dymerowska-Prokopczyk, S. Czerwinski & M. Switonski, 1998. Genes affecting meat and fat content in carcass of pig - recapitulation of Polish project "Pig genome mapping". Proc. Conf. Influence of genetic and non-genetic traits on carcass and meat quality of pigs. Pol. J.Food Nutr. Sci. 7/48, 4S: 70-76

Lahucky, R., L.L. Christian, L. Kovac, J. Stalder & M. Buerova, 1997. Meat quality assessed ante- and post- mortem by different ryanodine receptor gene status of pigs. Meat Sci. 43: 277-285

Rempel, W.E., Ming Yulu, J.R. Mickelson & C.F. Louis, 1995. The effect of skeletal muscle ryanodine receptor genotype on pig performance and carcass quality traits. Anim. Sci. 60: 249-257

Sellier, P. & G. Monin, 1994. Genetics of pig meat quality: A review. J. Muscle Foods, 5: 187-219

Wicke, M., S. Maak & G. van Lengerken, 1998. Structural and functional traits of the skeletal muscle for the improvement of pork quality. Proc. Conf. Influence of genetic and non-genetic traits on carcass and meat quality of pigs. Pol. J. Food Nutr. Sci. 7/48, 4S: 21-31

Zhang, W., D.L. Kuhlers & W.E. Rempel, 1992. Halothane gene and swine performance. J. Anim. Sci. 70: 1307-1313

Correlations between growth rate, slaughter yield and meat quality traits after the elimination of the RYR1 gene effect

W. Kapelanski[1], J. Kortz[2], J. Kuryl[3], T. Karamucki[2] & M. Bocian[1]*

[1]*University of Technology and Agriculture, Mazowiecka 28, 85-084 Bydgoszcz,* [2]*Agricultural University, Dr Judyma 24, 71-460 Szczecin,* [3]*Institute of Genetics and Animal Breeding, Polish Academy of Science, 05-551 Mrokow, Jastrzebiec, Poland*

Summary

Ninety pigs, 30 of each of the breeds Polish Landrace, Piétrain and Zlotniki Spotted were determined for the RYR1 genotype. There were 33 NN, 27 Nn and 30 nn. Meat quality traits were correlated with growth performance and killing-out yield. Besides the overall (r_o) correlations the intragroup (r_i) correlations within the genotype groups were calculated by covariance. The results indicated that the relationship between meat quality and growth rate was unrelated to the RYR1 genotype. The higher growth rate was followed by less red and more exudate meat. The relationship between meat quality and killing-out yield was affected by the RYR1 gene. Pigs with higher killing-out yield were of inferior meat quality.

Keywords: Pigs, Growth rate, RYR1 genotype, Meat quality

Introduction

Most reports on negative linkage between quantity and quality of meat concern the heavy muscled pigs (de Smet *et al.*, 1995; Garcia-Macias *et al.*, 1996; Rempel *et al.*, 1995), which are the most sensitive to stress. Effect of the halothane, or RYR1, gene on meat quality characteristics have been well documented (Cheah *et al.*, 1995; Kapelanski *et al.*, 1999b; Pommier & Houde, 1993; Sellier & Monin, 1994). However, the role of the n allele in performance traits improvements is not so obvious because differences for the three pig genotypes NN, Nn and nn in certain growth performance and carcass quality traits are not always significantly different (Fisher & Mellet, 1997; Hardge *et al.*, 1997; Kapelanski *et al.*, 1999a; Rempel *et al.*, 1995).

The aim of the study was to compare the overall and intragroup correlations between some performance and meat quality traits. This allows statistical elimination of the RYR1 gene effect and enables more precise inference about the real connection between the meat quality and productivity traits in pigs.

Material and methods

Thirty pigs of each breed group i.e., Polish Landrace (PL), Piétrain (P) and Zlotniki Spotted (ZS) with equal numbers of gilts and barrows in each group were tested from about 28 to 105 kilograms liveweight. The weight gain was checked every two weeks. Pigs were reared in groups of 10 pigs per pen and fed the standard diet *ad libitum*. Blood samples taken at exsanguination were analysed for the RYR1 genotype (Kuryl & Korwin-Kossakowska, 1993).

Meat quality traits were measured in the *longissimus lumborum* muscle. The pH_l value was measured using a pistol pH-meter and ultimate pH (pH_u) in meat-water slurry. The simplified spectrophotometric measurement of dominant wavelength, colour saturation and colour lightness was done on Spekol 11 with a reflectance attachment (Rozyczka *et al.*, 1968). Water holding capacity was determined according to the filter press method (Pohja & Niinivaara, 1957) and expressed as percentage of free water in meat. Meat plasticity was examined as outlined by Grajewska *et al.* (1998). Meat samples were analysed for pigment content and expressed as hematin (Hornsey, 1956).

From the immediately measured meat traits the meat quality criteria were calculated as Q (a synthetic value composed of eight meat characteristics) and I_2 (being a mathematical combination of pH_l and pH_u values) which were derived by Kortz (1986).

The pigs were divided into three groups according to their RYR1 genotype NN, Nn and nn. Data were statistically analysed using analysis of variance with respect to RYR1 genotype and sex. Additionally overall (r_o) and intragroup correlation coefficients (r_i) were calculated by covariance between the meat quality and performance traits. All statistical calculations were based on the methods of Snedecor (1956).

Results and discussion

In compared pigs the distribution of individuals in each of the RYR1 genotypes was similar, (33 NN, 27 Nn and 30 nn). Mean values of average daily gains (ADG), days on test (DAY) and killing-out yield (YLD) are presented in Table 1. Genotype effect for both growth traits (ADG and DAY) were not significant but for killing-out yield was significant in favour of the n allele (P < 0.01).

Table 1.   Growth rate and killing-out yield in pigs of three RYR1 genotype.

Traits	Genotype		
	NN	Nn	nn
Number, no	33	27	30
Average daily gain, g	719 ± 18	674 ± 21	706 ± 20
Day on test, days	108 ± 2.8	112 ± 3.2	109 ± 3.0
Killing-out yield, %	79.3 ± 0.3[aA]	80.2 ± 0.4[bA]	82.5 ± 0.6[cB]

a, b, c - Significant at P < 0.05;   A, B - Significant at P < 0.01

Many reports have shown that the RYR1 genotype or the gene with its linkage group significantly affected ADG, carcass lean content and YLD favourably and meat quality traits adversely (Garcia-Macias *et al.*, 1996; Rempel *et al.*, 1995; Zhang *et al.*, 1992). Our intention was to prove whether the relationship between the meat quality and performance traits exists then the RYR1 gene effect is eliminated.

The computed overall and intragroup correlation coefficients are presented in Table 2. As can be seen from the data the significant overall correlations of ADG and DAY with meat pigment content, dominant wavelength of meat colour and free water content in meat did not lose their significance after the elimination of the RYR1 gene effect (r_o and r_i of the same magnitude). The results indicated that higher ADG and shorter fattening period were associated with less red and more exudate meat irrespective of genotype of pigs.

On the other hand, the relationship between killing-out yield and most of the meat quality traits was affected mainly by the RYR1 genotype of pigs because intragroup correlation coefficients were not significant in comparison with overall correlation coefficients. From the above, the conclusion may be drawn that pigs with higher killing-out yield were characterised by inferior meat quality.

Table 2. *Overall (ro) and intragroup (ri) correlation coefficients between meat quality traits and growth rate and killing-out yield.*

Correlated traits	Av. daily gain		Days on test		Killing-out yield	
	r_o	r_i	r_o	r_i	r_o	r_i
Meat pigments, ppm	**-0.51**	**-0.51**	**0.51**	**0.51**	0.19	0.25
Dom. wavelength, nm	**-0.38**	**-0.37**	**0.39**	**0.41**	-0.10	-0.06
Colour saturation, %	-0.17	-0.18	0.18	0.20	**0.25**	0.05
Colour lightness, %	0.18	0.19	-0.18	-0.19	**0.37**	0.10
WHC, free water, %	**0.44**	**0.46**	**-0.49**	**-0.51**	**0.30**	0.12
Plasticity, cm^2	0.01	-0.01	0.02	0.03	**-0.58**	**-0.28**
pH$_1$	0.05	0.08	0.02	0.01	**-0.49**	-0.08
pH$_u$	0.04	0.03	-0.03	-0.04	-0.20	0.09
Q	-0.11	-0.02	0.13	0.14	**-0.48**	-0.18
I$_2$	0.05	0.10	0.02	-0.01	**-0.34**	0.09

Coefficients in bold type are statistically significant at P < 0.01

References

Cheah, K.S., A.M. Cheah & D.I. Kraugsgrill, 1995. Variation in meat quality in live halothane heterozygotes identified by biopsy samples of M. longissimus dorsi. Meat Sci. 39: 293-300

De Smet, S., H. Pauwels, I. Vervaeke, S. Demeyer, S. De Bie, W. Eeckhout & M. Casteels, 1995. Meat and carcass quality of heavy muscled Belgian slaughter pigs as influenced by halothane sensitivity and breed. Anim. Sci. 61: 109-114

Fisher, P. & F.D. Mellett, 1997. Halothane genotype and pork production. 1 Growth, carcass and meat quality characteristics. S. Afr. J. Anim. Sci. 27(1): 22-26

Garcia-Macias, J.A., M. Gispert, M.A. Oliver, A. Diestre, P. Alonso, A. Munoz-Luna, K. Siggens & D. Cuthbert-Heavens, 1996. The effects of cross, slaughter weight and halothane genotype on leanness and meat and fat quality in pig carcasses. Anim. Sci. 63: 487-496

Grajewska, S., W. Kapelanski & M. Bocian, 1998. Usefulness of meat plasticity measurements to assess the meat quality. Proc. Conf. Influence of genetic and non-genetic traits on carcass and meat quality of pigs. Pol. J. Food Nutr. Sci. 7/48, 4S: 141-144

Hardge, T., K. Kopke, G. Leuthold, K. Wimmers & Th. Paulke, 1997. Differences in allele frequencies of candidate genes for growth, carcass value and meat quality between extreme phenotypes of commercial pig breeds. 48th Annual Meeting of EAAP, August 25-28, Vienna

Hornsey, H.C., 1956. The colour of cooked cured pork. I. Estimation of the nitric oxide-haem pigments. J.Sci. Food Agric. 7: 534-540

Kapelanski, W., J. Kuryl, M. Bocian & B. Rak, 1999a. Effect of RYR1 gene on meat quality traits in Polish Landrace, Pietrain and Zlotniki Spotted pigs. Adv. Agric. Res. vol. 6, Fasc 2 (in press)

Kapelanski, W., J. Kuryl, M. Bocian & J. Kapelanska, 1999b. Effect of RYR1 gene on the growth rate and lean content in Polish Landrace, Pietrain and Zlotniki Spotted pigs. Adv. Agric. Res. vol. 6, Fasc 2 (in press)

Kortz, J., 1986. An attempt of appointment the synthetic index of pork meat quality as criterion for distinguish the normal, PSE and DFD muscle. Hab. Dissertation No 100. Ed. Agricultural Academy Szczecin

Kuryl, J. & A. Korwin-Kossakowska, 1993. Genotyping of HAL locus by PCR method explains some cases of incomplete penetrance of Hal^n gene. Anim. Sci. Papers and Rep. 11: 271-277

Pohja, M.S. & F.P. Niinivaara, 1957. Die Bestimmung der Wasserbindung des Fleisches mittels der Konstantdruckmethode. Fleischwirtschaft, 9: 193-196

Pommier, S.A. & A. Houde, 1993. Effect of the genotype for malignant hyperthermia as determined by a restriction endonuclease assay on the quality characteristics of commercial pork loins. J. Anim. Sci. 71: 420-425

Rempel, W.E., Ming Yulu, J.R. Mickelson & C.F. Louis, 1995. The effect of skeletal muscle ryanodine receptor genotype on pig performance and carcass quality traits. Anim. Sci. 60: 249-257

Rozyczka, J., J. Kortz & S. Grajewska-Kolaczyk, 1968. A simplified method of the objective measurement of colour in fresh pork meat. Rocz. Nauk Rol. 90-B-3: 345-353

Sellier, P. & G. Monin, 1994. Genetics of pig meat quality: A review. J. Muscle Foods, 5: 187-219

Snedecor, G.W., 1956. Statistical Methods. 5-th ed. Ames, Iowa. The Iowa State College Press

Zhang, W., D.L. Kuhlers & W.E. Rempel, 1992. Halothane gene and swine performance. J. Anim. Sci. 70: 1307-1313

Effect of the RYR 1 gene on meat quality in pigs of Large White, Landrace and Czech Meat Pig breeds

R. Bečková & P. David

Research Institute of Animal Production in Prague - Uhrineves, Department of Pig Nutrition and Meat Quality in Kostelec n.O. Czech Republic

Summary

From the meat traits we measured the values of pH_1 and pH_{24} in the LD muscle, the content of intramuscular fat (IF) and the percentage of lean meat (by a two point method). Genotypes of RYR 1 were determined by DNA test and frequencies were as follows: Large White (BU) 26 N/N and 8 N/n, Landrace (L) 20 N/N and 8 N/n, Czech Meat Pig (CVM) 13 N/N, 21 N/n and 9 n/n. The differences in BU pigs between N/N and N/n genotypes were not significant. PSE meat has been found in the N/n genotype only (5.88 %). The differences in L pigs between N/N and N/n genotypes were not significant again but more pronounced, except pH_1 ($P < 0.05$). The occurrence of PSE meat was much higher (17.8 %, N/N - 80 % and N/n 20 %). Significant differences between N/N, N/n and n/n genotypes were found in meat traits in the CVM breed, where all three genotypes were found: pH_1 ($P < 0.05$), IF (N/N and n/n - $P < 0.05$); the percentage of lean meat (N/N and n/n - $P < 0.05$). The frequencies of genotypes correspond to the occurrence of PSE meat in this breed 39.5 % (N/N - 0 %, N/n - 58.8 % and n/n - 41.2 %). The results obtained proved that the pigs of N/N genotype showed a higher resistance to stress, performance of PSE meat, higher content of IF, and a lower percentage of lean meat. The heterozygous N/n genotype showed this meat traits to be medium.

Keywords: Genotypes of RYR, pH, Intramuscular fat, Lean meat

Introduction

Meat quality, despite the fact that we consider it a quantitative characteristic, shows considerable dependence on stress susceptibility. Decisive influence on pig meat quality in population has the frequency of the halothane genotypes. Because of intensified selection of pigs with a higher proportion of muscles, the partial deterioration of meat quality has occurred (PSE, DFD meat). By decreasing the proportion of fat tissue, the content of intramuscular fat also decreased. The changes in meat quality show that animals with a high ability to produce protein and with a characteristic musculature, have a genetic predisposition to high stress susceptibility. Stress susceptibility of pigs is genetically given by the RYR gene that is located at the site of the assumed halothane gene.

Eikelenboom *et al.* (1976) state that halothane reagents have better carcass value but worse meat quality. Webb *et al.* (1986) found lower feed consumption, larger area of *ml.l.t.* and ham v. lean meat ratio and worse characteristics of meat quality in halothane positive animals. Bečková *et al.* (1989) engaged in an analysis of relationships between indices of slaughter value and physical indices of meat quality. Sather *et al.* (1996) & Zhang *et al.* (1992) state the influence of the RYR 1 genotype determined by DNA test on growth and carcass quality. Significantly higher weight gain and higher percentage of lean meat in halothane positive animals compared with halothane negative animals was found by Wittman *et al.* (1985). In contrast, lower weight gain and fat thickness in animals with the n/n genotype was found by Matthes & Schwerin (1995). Higher disposition to PSE meat production, less intramuscular fat, greater consumption of food and lower intensity of growth in N/N pigs but greater muscle deepness was found by Sather *et al.* (1996). Also Park *et al.* (1998) states that n/n genotype pigs have significantly lower weight gain and lower growing ability compared to N/N genotype pigs. In the CR the systematic selection on stress disposition has been introduced since 1994 and 90 % of the breeding herds are already mapped.

The goal of the present work was focused on finding the influence of the RYR 1 genotypes on the quality of dams of the Large White and Landrace breeds raised in the CR and the sires of the Czech Meat pig breed.

Material and Methods

The experiment was carried out under standard conditions at the stations of fattening and carcass value. In the investigations were included dam breeds of the Large White (BU) and the Landrace (L) of Czech origin (34 and 28 animals respectively), and 43 animals of the sire breed CVM (Czech Meat Breed). Within the individual breeds the equal representation of sex was respected and tested pairs had to have originated from several boars of different genotype lines. Of the meat quality characteristics we investigated the values of pH_1 (45 minutes *post mortem*), pH_{24} (24 hours *post mortem*) the content of intramuscular fat (IF) and percentage of musculature according to respective methodologies. The genotypes of the RYR 1 gene were determined by DNA test in the Institute of Genetics (MZLU in Brno). The results were assessed by the common statistical methods and dispersion analysis.

Results and Discussion

In the investigation 34 pigs in total were included, 17 gilt's and 17 barrows of the BU breed with average live weight x = 101.6 ± 1.9 and 101.6 ± 2.0 kilograms respectively. The average daily gain in test was found only with pairs (gilt and barrow) and reached x = 961.8 ± 72.4 g with feed consumption x = 35.1 ± 2.3 MJ of metabolizable energy (ME) per 1 kilogram of gain. The L breed was represented by 28 pigs (14 gilt's, 14 barrows) with the average live weight x = 101.6 ± 2.3 and 101.7 ± 1,7 respectively. The average daily gain in test was presented as x = 978.9 ± 81.2 g. ME intake per kilogram of daily gain amounted to x = 34.8 ± 2.3 MJ.

Table 1 presents the influence of genotype on meat quality in the BU breed expressed by pH_1, pH_{24} characteristics, content of intramuscular fat and percentage of muscle tissue. In tested pigs only N/N genotypes - 26 animals (76.5 %) and N/n - 8 animals (23.5 %) were found. Investigated results correspond with the frequency of genotypes determined by Kahánková (1998) in BU breed piglets (77.8 % and 18.5 % respectively). Even though the differences in the values of investigated characteristics between N/N and n/n are evident they are not statistically significant.

Table 1. *Influence of the RYR 1 gene on characteristics of meat quality (BU).*

Characteristic	Genotype	n	x	±	sx	v	Significance
PH1	N/N	26	6.31	±	0.24	3.84	
	N/n	8	6.18	±	0.26	4.28	-
pH24	N/N	26	5.54	±	0.18	3.19	
	N/n	8	5.49	±	0.145	2.659	-
IF (g.kg-1)	N/N	26	18.18	±	3.287	18.593	
	N/n	8	17.679	±	1.362	7.491	-
% lean	N/N	26	55.500	±	3.467	6.247	
meat	N/n	8	55.625	±	2.326	4.182	-

+ $P < 0.05$ pH_1 - 45 min *post mortem*, pH_{24} hours *post mortem*, IF - intramuscular fat, BU – Large White

The level of intramuscular fat (IF) within the investigated groups is low especially in N/n genotype pigs. The determined values of pH_1 and pH_{24} correspond with the occurrence of quality deviance of PSE type meat - 5.9 %, which was recorded only within the N/n genotype (25 % of the total number of heterozygote animals). Proportion of muscle within both investigated genotypes is on average nearly at the same level (55.5 and 55.6 % respectively). In the L breed where there is an assumption of a higher frequency of "n" all the differences between the N/N and n/n genotypes are already more evident except the pH_1 characteristic, where significant differences between the genotypes were found, the results are again statistically no significant (Table 2).

The frequency of genotypes within the groups of pigs of the L breed N/N - 71.4 %, N/n - 28.6 % are rather different compared to the figures given by Hradil (1985) (N/N - 60.5 %, N/n - 36.1 % and n/n - 3.4 %) Within the investigated sample the n/n genotype was not found again.

Table 2. *Influence of the RYR 1 gene on characteristics of meat quality (L).*

Characteristic	Genotype	n	x	±	sx	v	Significance
pH_1	N/N	20	6.14	±	0.17	2.82	
	N/n	8	5.97	±	0.25	4.12	+
pH_{24}	N/N	20	5.50	±	0.19	3.42	
	N/n	8	5.49	±	0.18	3.35	-
IF $(g.g^{-1})$	N/N	20	18.99	±	4.87	25.65	
	N/n	8	18.03	±	6.14	34.08	-
% lean	N/N	20	55.50	±	3.47	4.18	
meat	N/n	8	56.25	±	2.33	6.25	-

+ $P < 0.05$; L - Landrace

The presence of PSE meat in individual animals of this breed was much higher (17.8 %). 80 % of the quality deviation was revealed in the N/n genotype and the remaining 20 % in the N/N genotype. All the animals of the N/N genotype in which the PSE meat was found showed high lean meat value (58 - 62 % of muscle). Dvořák (1997) states that even in pigs with the N/N genotype the PSE meat can appear after slaughter in connection with high meat value and an intensive impact of stress factors.

Intramuscular fat (IF) content with both genotypes again doesn't reach required optimum values (about 25.00 $g.kg^{-1}$) (Bejerholm *et al.*, 1986). Within this group of L breed pigs the individuals of N/n genotype reached higher values of meatiness by nearly 1 %, but the difference is again statistically non significant (x = 55.5 resp. 56.3 %). Within both mentioned breeds only the pigs of N/N genotype reached meatiness in some cases of more than 60 % of muscle, but the pigs of N/n genotype, the heterozygotes, were mainly within the values of 55.0 - 59.9 % muscle.

Sire CVM breed represented by 43 animals reached in test average daily gain x = 802.7 ± 71.7 g with consumption of ME x = 35.5 MJ per 1 kilo of weight gain. Influence of genotype on meat quality in this breed is shown in table 3. Within this breed all three genotypes were found. The frequency of genotypes was as follows: N/N 13 - 30.2 %; N/n 21 - 48.8 %; n/n 9 - 20.9 %. Similar results were presented by Dvořák (1997) who found the genotype frequency N/N - 25 %, N/n - 50 %, n/n - 25 % in the CVM breed. From presented results it is clear that in the CVM breed, except for pH_{24} there are in all investigated characteristics of meat quality significant differences between N/N and n/n genotypes and in the case of pH_1 also between the N/n and n/n genotypes. The low values of pH_1 in the N/n and n/n genotypes corresponded well to the presence of quality deviation of PSE type that was found in 17 pigs (39.5 %), of that at N/N genotype 0, N/n - 10 (47.6 %), n/n - 7 (77.7 %). Determined IF values in the N/N genotype were very good (x = 25.9 ± 7.8), in the N/n genotype they dropped but had the lowest variability in the N/N genotype where they were already low (x = 19.6 ± 6.4). Schwörer (1989) states that pigs of LW, Landrace and Piétrain breeds have a lower proportion of IF. The Hampshire breed takes middle position and the Duroc breed has the highest IF. He also states that besides the breed correspondence the IF is influenced by the genotype of animal in the halothane test. The lower proportion of IF have pigs of n/n genotype. In contrast, in the case of lean meat the highest percentage of muscle was reached by the individuals of n/n genotype and N/n genotype had nearly the same level as N/N. Mathes et al. (1995) dealt with the relationship of genotypes to percentage of lean meat in meat breeds. The highest percentage of lean meat he also found in the n/n genotype, the middle in N/n and the lowest in N/N.

Table 3. Influence of the RYR 1 gene on characteristics of meat quality (CVM).

Characteristic	Genotype	n	x	±	sx	v	Significance	
pH_1	N/N	13	6.24	±	0.25	4.15	1 : 2	+
	N/n	21	5.93	±	0.30	4.78	2 : 3	+
	n/n	9	5.71	±	0.12	3.97	1 : 3	+
pH_{24}	N/N	13	5.53	±	0.20	3.56		
	N/n	21	5.47	±	0.21	3.64		-
		9	5.44	±	0.18	3.12		
IF $(g.kg^{-1})$	N/N	13	25.83	±	7.75	30.01	1 : 3	+
	N/n	21	22.04	±	4.97	20.72		
		9	19.56	±	6.37	32.34		
% lean	N/N	13	55.40	±	2.50	4.52	1 : 3	+
meat	N/n	21	55.95	±	3.20	5.784		
		9	56.56	±	3.01	5.313		

+ P < 0.05; CVM – Czech Meat Pig

Finally we can state that in investigated groups of pigs of BU and L breeds the n/n genotype was not found. We attribute this absence to the higher resistance of dam breeds to stress and systematical selection on stress disposition carried out in breeding herds. Despite not finding significant differences in characteristics of pig meat between N/N and N/n genotypes from the results, it is evident that N/N genotype pigs compared to N/n genotype have higher stress resistance and subsequently lower occurrence of the quality deviation of PSE type. In the sire CVM breed that presented all three types of genotypes, the differences between the characteristics in the respective genotypes are more significant between N/N and n/n genotypes at the level of significance $P < 0.05$. The genotype n/n presents lower pH values, higher percentage of PSE meat, lower content of intramuscular fat and higher percentage of muscle compared to the N/N genotype.

References

Bečková, R., 1989. Influence of genetic and intravital factors on meat quality. Final Report, VÚCHP Kostelec nad Orlicí

Bejerholm, C., 1986. Effect of intramuscular fat level on eating quality of pig meat. 32. Europ Congress of Meat Research Workers

Dvořák, J., 1997. Results of DNA test in pigs, Institute of Genetic MZLU, Brno, 37

Eikelenboom, G. & D. Minkema, 1976. The application of Halothane-Test. Differences in Production Characteristics between Pigs Qualified as Reactors (MHS-Susceptible) and Non-Reactors. Zeit, 1-5

Hradil, R., 1998. The analysis of genome of farm animals by the PCR method. Dissertation paper, Brno, 118

Kahánková, L., 1998. Variation of pig population from the point of genetic markers and their relation to performance. Dissertation paper, Brno, 135

Matthes, W. & M. Schwerin, 1995. Use of the malignant hyperthermia syndrom gene test for the determination of stress susceptibility. Neue Landwirtschaft 6: 56-58

Park, Y. I. et al., 1998. Effect of porcine stress syndrome genotype on performance traits in swine. 6th World Congress on Genetics Applied to Livestock Production, Australia, 23: 668

Sather, A..P. & S.D.M. Jones, 1996. The effect of the halothane gene on pork production and meat quality of pigs reared under commercial conditions. Canad. J. of Anim. Sci. 76: 507-516

Schwörer, D., 1989. Genetic variation in intramuscular fat content and sensory properties of pork. EAAP, Dublin

Webb, A. J. & S.P. Simpson, 1986. Performance of British Landrace pigs selected for high and low incidence of halothane sensitivity. 2. Growth and carcass Traits. Animal Production 43: 493-503

Wittman, M., 1985. Meat quality in pigs sensible on stress. International Agricultural Magazine, 29: 77-79

Zhang, W., Kuhlers, D.L. & W.E. Rempel, 1992. Halothane gene and swine performance, of Anim. Sci. 70: 1307-1313

Development of a highly accurate DNA-test for the *RN* gene in the pig

C. Looft[1], D. Milan[2], J.T. Jeon[3], S. Paul[1], C. Rogel-Gaillard[4], V. Rey[2], A. Tornsten[3], N. Reinsch[1], M. Yerle[2], V. Amarger[3], A. Robic[2], E. Kalm[1], P. Chardon[4] & L. Andersson[3]

[1] *Institute of Animal Breeding and Husbandry, Christian-Albrechts-University, D-24098 Kiel, Germany*
[2] *Laboratoire de Génétique Cellulaire, INRA, F-31326 Castanet Tolosan, France*
[3] *Department of Animal Breeding and Genetics, Swedish University of Agricultural Sciences, S-751 24 Uppsala, Sweden*
[4] *Laboratoire de Radiobiologie et d'Etude du Génome, INRA/CEA, F-78352 Jouy-en-Josas Cedex, France*

Summary

The porcine *RN* gene, previously mapped to chromosome 15, has a large effect on meat quality. Four research groups have now joined their efforts in an attempt to identify the *RN* gene. Reference families comprising altogether about 1000 backcross animals were collected for precise linkage mapping of *RN*. The map shows that *RN* is located in a short interval between microsatellites Sw2053 and Sw936. FISH mapping of YACs containing Sw2053 or Sw936 allowed a regional assignment to chromosome 15q2.5. Pig markers were developed for nine genes assigned to the corresponding region in humans and mapped using a pig radiation hybrid (RH) panel. A contig of about 2 Mbp containing the *RN* gene was constructed by screening YAC (yeast artificial chromosomes) and BAC (bacterial artificial chromosomes) libraries. Additional markers were isolated by sequencing BAC ends. With these resources we developed markers in almost total linkage disequilibrium with the *RN* allele in various populations. They are now available and can be used in Marker Assisted Selection programmes.

Keywords: Pigs, Meat quality, RN gene

Identification of the RN gene

In 1985 it was shown that meat from Hampshire animals is characterised by a low ultimate pH-value, a high glycogen content and a reduced technological yield in cured cooked ham processing. As this phenomenon occurred only in the Hampshire breed, it has been called the "Hampshire effect (Monin & Sellier, 1985). Complex segregation analysis by Le Roy *et al.* (1990) revealed the existence of a major gene with one dominant allele, the *RN⁻* allele, with an unfavourable effect on the technological yield and one normal recessive allele, the *rn⁺* allele.

The *RN* genotype can be derived from glycogen measurements. But there is a necessity to develop a DNA test for the gene, because the determination of the genotype by physiological analysis has some inaccuracies. Furthermore it is only possible after slaughtering or - in live animals - using muscle biopsies that are in contradiction to animal welfare regulations in some countries.

Genetic Mapping

In order to develop such a test three groups in Sweden, France and Germany mapped the *RN* gene. After the establishment of resource families by crossing Hampshire pigs with white lines and crossing animals of a synthetic line, determination of the animals *RN* genotype, and performing a genome screen with microsatellite markers and linkage analysis showed the location of the *RN* gene on chromosome 15 in the interval Sw120-Sw936 (figure 1; Milan *et al.*, 1996; Mariani *et al.*, 1996; Reinsch *et al.*, 1997) and on the cytogenetic map in the 15q25 region by fluorescence in situ hybridisation (FISH) with YAC clones from the RN-region.

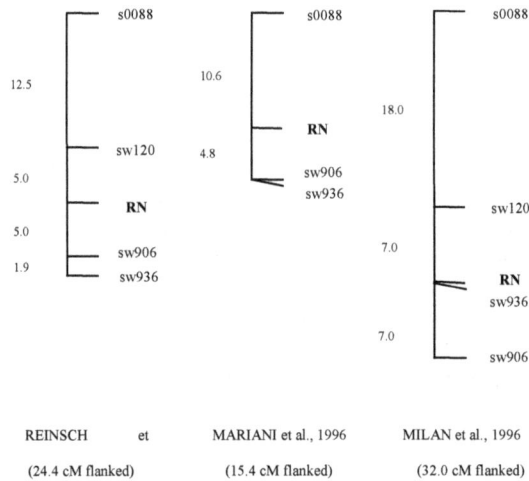

Figure 1. Comparison of male multipoint map details in the RN-region from three studies.

In the next phase comparative mapping information was used to determine the orthologous region of the *RN*-locus on human chromosome 2 in the region 2q33-q36 (Robic *et al.*, 1999). Furthermore candidate genes like *UGP2* and *PPP1R3* were defined and excluded because of the results of mapping experiments (Milan *et al.*, 1996; Looft *et al.*, 1996).

Positional cloning

In 1997 four research groups from the Institut Nationale de la Recherche Agronomique (France), Swedish University of Agricultural Sciences (Sweden) and the Christian-Albrechts-University (Germany) initiated a collaborative project to construct a high density linkage map in the *RN*-region, identify the *RN* gene and develop a DNA-test.

    The applied strategy is shown in figure 2. In a first step additional markers were isolated in the target region by applying a "trans-species shuttling□ strategy as proposed by Georges & Andersson (1996).

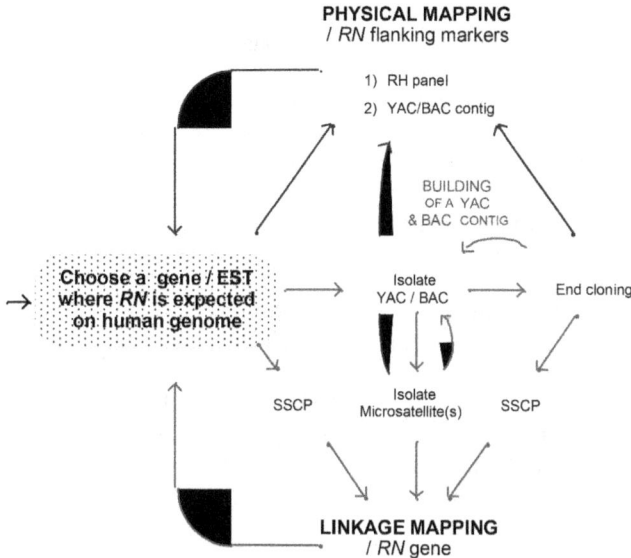

PHYSICAL MAPPING
/ *RN* flanking markers

1) RH panel
2) YAC/BAC contig

BUILDING
OF A YAC
& BAC CONTIG

Choose a gene / EST
where *RN* is expected
on human genome

Isolate
YAC / BAC

End cloning

SSCP

Isolate
Microsatellite(s)

SSCP

LINKAGE MAPPING
/ *RN* gene

Figure 2. Strategy to develop markers in total linkage disequilibrium to the RN gene.

Human and murine sequence information have been used to develop markers in the orthologous region of the porcine genome and Robic *et al.* (1999) established a radiation hybrid map of the *RN*-region, which demonstrated a conserved gene order with the human and mouse genome. These results clearly show that no candidate gene for *RN* has yet been mapped to the corresponding region in humans and mice.

Furthermore the isolated markers have been used to detect polymorphisms and to isolate BAC and YAC clones. End-sequencing of BAC clones allowed us to build up a BAC contig of about 2 Mbp. In parallel to the construction of the BAC contig isolated polymorphic markers were used to genotype the three resource families with more than 1000 informative meioses for the *RN* gene. Linkage analysis showed that the *RN* gene was mapped into an interval of less than 1 cM. The isolated markers are in almost total linkage disequilibrium with the *RN⁻* allele in various populations and can be used in Marker Assisted Selection programmes with a high accuracy. BAC clones containing the *RN* gene are now used for direct sequencing with the aim of identifying all transcripts from the actual region and evaluate them as potential candidate genes.

References

Georges, M. & L. Andersson, 1996. Livestock genomics comes of age. Genome. Res. 6: 907-921

Le Roy, P., J. Naveau, J.M. Elsen & P. Sellier, 1990. Evidence for a new major gene influencing meat quality in pigs. Genetic Research Cambridge 55: 33-40

Looft, Chr., S. Paul, B. Brenig & E. Kalm, 1996. Cloning and sequencing of the porcine UDP-glucose pyrophosphorylase gene - a candidate gene for the RN-locus. Anim. Genet. 27, Suppl. 2: 81

Mariani, P., K. Lundström, U. Gustafsson, A.-C. Entfält, R.K. Juneja & L. Andersson, 1996. A major locus (RN) affecting muscle glycogen content is located on pig chromosome 15. Mamm. Genome 7: 52-54

Milan, D., N. Woloszyn, M. Yerle, P. Le Roy, M. Bonnet, J. Riquet, Y. Lahbib-Mansais, J.-C. Caritez, A. Robic, P. Sellier, J.-M. Elsen & J. Gellin, 1996. Accurate mapping of the "acid meat" RN gene on genetic and physical maps of pig chromosome 15. Mamm. Genome 7: 47-51

Monin, G. & P. Sellier, 1985. Pork of low technological quality with a normal rate of muscle pH fall in the immediate post-mortem period: the case of the Hampshire breed. Meat Science 13: 49-63

Reinsch, N., Chr. Looft, I. Rudat & E. Kalm, 1997. The Kiel RN-project - final porcine chromosome 15 mapping results. Journal of Animal Breeding and Genetics 114: 133-142

Robic, A., V. Seroude, J.T. Jeon, M. Yerle, L. Wasungu, L. Andersson, J. Gellin & D. Milan, 1999. A radiation hybrid map of the RN region in pigs demonstrates conserved gene order compared with the human and mouse genomes. Mamm. Genome 10: 565-568

Performances of the Piétrain ReHal, the new stress negative Piétrain line

P. L. Leroy & V. Verleyen[1]

[1] *Department of Genetics, Faculty Veterinary Medicine B43, University of Liège, B-4000 Liège Belgium.*

Summary

Piétrain ReHal was created by introgressing the negative stress gene from the Large White into Piétrain (successive back-cross (BC)).

The pigs born in station from Piétrain ReHal boars used as terminal boars are characterised, on average, by a feed efficiency of 2.959, a daily gain during the fattening period of 649 grams, a killing out percentage of 82.6 percent with an SKG2 meat percent estimation of 58.55% and a back fat thickness of 2.005 cm.

A total of 5,002 piglets were obtained from commercial sows and boars of different genetic origin, and Piétrain ReHal (Nn) boars were compared to Landrace and Piétrain pure-bred animals. All the animals were born on 2 farms and fattened on 19 farms. Meat percentage was analysed with a linear fixed model including fattening farm effect, sow line, sex, boar within breed and breed of the boar. Results indicated that Piétrain ReHal heterozygote boars performed quite well (meat %=58.93%) and that the estimated meat percentage is close to the pure Piétrain results (59.48%) and better than Landrace boars (57.99%).

In 1997 and 1998, BC5, BC6 and BC7 (=255/256 Piétrain) generations have been produced. More recently, it has been decided to produce the Piétrain ReHal homozygote stress negatives called Piétrain ReHal[cc]. A large number of Piétrain ReHal[cc], with the meat and carcass performance of Piétrain pigs, are produced now in Belgium.

Keywords: Piétrain, Stress negative, Meat quality, Halothane locus

Introduction

The development of a new stress negative Piétrain line was launched at the Faculty of Veterinary Medicine of the University of Liège in the 1980's. Two main objectives of the project were defined.

Firstly, fundamental research was required on the Halothane locus and its impact on meat production and quality, especially in a specific genetic background like the Piétrain which is completely different from other halothane positive lines (Nezer *et al.*,1999).

Secondly, there was a increasing requirement for pigs that were completely free of the stress gene, both to reduce carcass loss due to unacceptable pH levels and PSE-PSS syndrome and to meet consumer requirements for a reduction in pre-medication before transport.

The Piétrain ReHal (resistant to halothane anaesthetic gas) was created by Hanset and co-workers (Hanset *et al.*,1989) by introgressing the negative stress gene (N) from the Large White into the Piétrain. The background Piétrain genome has been recovered by successive backcrossing. The first backcross generation (BC1) were 75% Piétrain, and at the present moment the seventh backcross (BC7 or 255/256 Piétrain) has been reached.

A total of 600 pigs, stress negative animals, heterozygotes (Nn) that are 99.6% Piétrain and also homozygotes (NN), born to Nn parents, constitute the foundation sire line of the research station farm at the campus of Sart Tilman (University of Liège). The stress negative heterozygotes animals (Nn) represent 47% and the stress negative homozygotes (nn) 13% of the total number of pigs born in 1999.

The used of Piétrain ReHal boars in station

Three commercial sow lines were obtained from a private company (Detry SA) and transferred in November 1995 to the University station as part of a project co-financed by Detry SA and the Federal Ministry of Agriculture.

The objective of the study was to estimate growth and carcass performance of Piétrain ReHal (Nn) boars on commercial sows under station conditions.

The 3 sow lines were crossed with Piétrain ReHal boars. All the animals received the same feed composition with the same management conditions.

An integer value (from 11(best) to 17 (worst)), considered as a subjective ranking of the carcass, and an estimated meat percentage (obtained from a camera giving the SKG2 meat percentage estimation) were available from the system currently used by the private slaughterhouse participating in the study.

The growth and carcass performances of the total data set (83 NN, 188 Nn, 72 nn and 37 with unknown halothane genotype) indicate clearly that the reconstruction of the Piétrain breed has been almost attained (table 1). The Europ ranking distribution shows that more than 85% of the pigs are in the S and E highest classes (table 2). The pigs born from Piétrain ReHal boars used as terminal boars are also characterised, on average, by a feed efficiency of 2.959, a daily gain during the fattening period of 649 grams, a killing out percentage of 82.6 percent with an SKG2 meat percent estimation of 58.55 and a back fat thickness of 2.005 centimetres. A linear fixed model including sow line, genotype (NN, Nn, nn) and sex does not explain the differences in weaning weight. The same model including weaning weight as a covariate explains 12.6% of the variation in slaughter weight. The effect of genotype on slaughter weight corrected for weaning weight was not significant. A third fixed linear model including sow line (L, F), genotype (NN, Nn, nn), sex fixed effects and weight at weaning and weight at slaughter as a covariable explained from 6.3 to 18.1 of the variation in carcass traits.

Table 1.   Growth and carcass performance of pigs obtained from Piétrain Piétrain ReHal boars at the Research Station farm of the University of Liège (380 animals).

Variable	Mean	S
Birth weight (g)	1592.2	427.8
Weaning weight (Kg)	7.69	1.66
Slaughter weight (Kg)	114.5	14.4
Subjective ranking (11-17)	14.2	1.1
Carcass length (cm)	83.3	3.4
Carcass weight (Kg)	95.9	9.8
Killing out %	82.6	50.8
Meat%	58.6	3.4
Back fat thickness (cm)	2.0	0.6
pH loin after 1 h.	6.3	0.3
pH loin after 24 h.	5.7	0.2

The use of Piétrain ReHal heterozygote (Nn) boars on commercial sow lines in Farm conditions

In parallel with the study of the Piétrain ReHal on station, where Piétrain ReHal animals have been continuously produced from heterozygote Piétrain ReHal (Nn) and new Piétrain (nn) sows, another experiment was set up in order to test the Piétrain ReHal line under commercial conditions. 5,002 piglets were obtained from sows and boars of different genetic origin; Piétrain ReHal (Nn) boars were compared to Landrace and Piétrain purebred animals. All the animals were born on two farms and fattened on 19 farms.

Table 2. The use of Piétrain Piétrain ReHal boars on station. Europ carcass ranking of pigs (n=343) born from different commercial sow lines crossed with Piétrain ReHal boars.

Europ Rank.	N	%
S	123	35.9
E	171	49.9
U	48	14
R	1	0.3

The data were analysed by a linear fixed model including: fattening farm effect, sow line, sex, boar within breed and breed of the boar. The distribution of the data by origin of boars is given in table 3. The descriptive statistics concerning some growth and carcass traits are given in table 4. The overall results (table 4) are influenced by the frequency of Piétrain boars in comparison with other origins. The percentage of variation explained by the linear model ranged from 10.1% to 21.9% and the relative importance of each effect is illustrated in figure 1. Differences between breed of boars were significant and are given for subjective ranking (11-17) and percentage of meat in table 3.

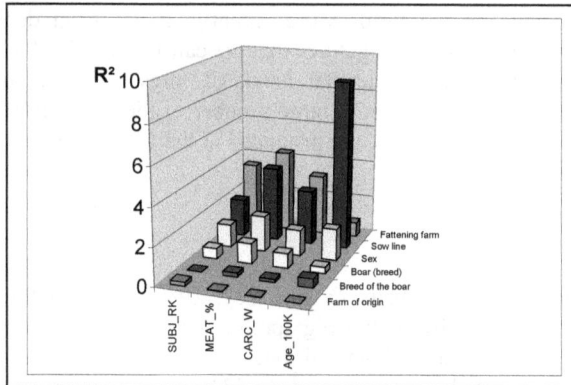

Figure 1. The use of Piétrain ReHal boars (farm conditions). Percentage of the variation (R²%) of subjective ranking (SUBJ_RK), percentage of meat (MEAT%), carcass weight (CARC_W) and age at 100 kilos (AGE_100K) explained by the different effects included in the linear model.

The results of the subjective ranking and meat percentage indicate that Piétrain ReHal heterozygote boars perform quite well and that the estimated meat percentages are close to the Piétrain results. Piétrain ReHal boars give better results than Landrace boars.

Table 3. *Frequency distribution of Piétrain ReHal boars used on commercial sow lines and Least square means of subjective ranking and estimated meat % (farm conditions, n=5,008).*

Origin of the boars	Frequency	%	Subject. Rank.	Meat % estim.
Piétrain (Belgium)	3,912	78.1	13.78	59.48
Piétrain (France)	203	4.1	13.93	58.99
Piétrain ReHal	582	11.6	13.99	58.93
Landrace (Belgium)	191	3.8	14.08	57.99
Landrace (France)	120	2.4	14.42	56.95

Table 4. *The use of Piétrain ReHal boars (farm conditions). Growth and carcass performances of 5,008 pigs born from commercial sow lines and from different boar type.*

Variable	Mean	S
Age at weaning (days)	27.7	3
Age at slaughter (days)	213.2	18.8
Subjective ranking (11-17)	13.9	1.1
Meat%	59.4	3.5
Carcass weight (Kg)	92.5	10.4
Slaughter weight (Kg)	112.8	12.7
Age at 100 kg (days)	209.5	18.5

The Piétrain ReHal[CC], the Piétrain ReHal homozygote stress negative

In 1997 and 1998 BC5, BC6 and BC7 generations have been produced. For each year, litter number per year was on average 2.06 and the number of piglets born and weaned per sow per year were respectively 19.02 and 15.45. Better results have been obtained due to improved management conditions on the farm. More recently, it has been decided to produce the Piétrain ReHal homozygote stress negatives that we call Piétrain ReHal[cc]. These homozygote stress negative pigs are, at the molecular level, CC pigs and they correspond to the NN nomenclature of the previous studies. A large number of Piétrain ReHal[cc], with the meat and carcass performance of Piétrain pigs, are produced at the University Station farm and also by Belgian Piétrain breeders.

References

Hanset, R., C. Dasnois, C. Michaux & P. Leroy, 1989. Looking for individual genes of muscle hypertrophy in the pig: preliminary analysis of a F2 of the Piétrain x Large-White cross. 40th annual meeting of the European Association for Animal Production E.A.A.P. Dublin, Vol. 1, G2.9, 71-72

Leroy, P.L, V. Verleyen & J-P. Detry, 1999. Le porc Piétrain résistant au stress (ReHal) dans la filière porcine. 4ème Carrefour des productions animales de Gembloux. Edited by the Faculté des Sciences Agronomiques, B-5030 Gembloux, Belgium, 39-40

Nezer, C., L. Moreau, B. Brouwers, W. Coopieters, J. Detilleux, R. Hanset, L. Karim, A. Kvasz, P. Leroy & M. Georges, 1999. An imprinted QTL with major effect on muscle mass and fat deposition maps to IGF2 locus in pigs. Nature Genetics 21: 155-156

Effect of the RN⁻ gene on the growth rate carcass and meat quality in crossbreeding of Large White sows with P-76 boars

M. Koćwin-Podsiadła, W. Przybylski, E. Krzęcio, A. Zybert & S. Kaczorek

Pig Breeding Department, Faculty of Agriculture
University of Podlasie, 08-110 Siedlce, 14 Prusa Str., Poland

Summary

The aim of study was to analyse the effect of the RN⁻ gene on growth rate, carcass and meat quality in pigs with RN⁻rn⁺ and rn⁺rn⁺ genotypes derived from crossbreeding of Polish Large White sows with P-76 boars. The investigations covered 50 animals (28 castrated males and 22 gilts) - 25 RN⁻rn⁺ and 25 rn⁺rn⁺ that originated from crossing 16 Polish Large White breed sows with 3 boars P-76 French line*. A positive effect of the RN⁻ allele for slaughter age and back fat thickness was noted. Pigs with genotypes RN⁻rn⁺ were 20 days younger at slaughter than rn⁺rn⁺ and also smaller average back fat thickness (at the level of significance $P \le 0.07$). Results of the present investigation showed a negative effect of the RN⁻ gene on meat quality. Pigs with RN⁻rn⁺ genotype had: lower pH_{24}, more pale meat and higher glycolytic potential evaluated in biopsy samples of *Longissimus dorsi* muscle. They also had lower technological yield of meat in cooking and smoke-curing, processing about 10% and 5% respectively in comparison to rn⁺rn⁺ pigs.

Keywords: RN gene, Meatiness, Meat quality, Technological yield

Introduction

Unfavourable phenotypic influence of the RN⁻ allele on meat quality traits manifests itself with an increase of glycolytic potential in white muscles such as *Longissimus dorsi* and *Semimembranosus* and meat lightness, a decrease in pH_u, a reduction of protein content in muscle tissue and about an 8% decrease of technological yield in cooking and curing processing (Monin 1994, Przybylski *et al.*, 1996). A small but unfavourable influence of the RN⁻ gene on growth rate and carcass quality is also observed (Le Roy *et al.*, 1995; Lundström *et al.*, 1994). Moreover, meat from carriers of this gene has better taste and aroma and needs less shear force in the slicing process (Lundström *et al.*, 1994; Monin 1994). Milan *et al.* (1995) discovered RN locus on chromosome 15, 18cM from S0088 marker.

The aim of this study was to analyse the effect of the RN⁻ gene on growth rate, carcass and meat quality and also technological yield of meat of pigs with RN⁻rn⁺ and rn⁺rn⁺ genotypes derived from crossbreeding of Polish Large White sows with P-76 boars.

---

* Boars originated from specially produced herd with RN⁻ gene only for scientific researches (from Pen Ar Lan Breeding Company)

Material and methods

The investigations were carried out on 50 animals (28 castrated males and 22 gilts) - 25 RN⁻ rn⁺ and 25 rn⁺rn⁺ that originated from crossing 16 Polish Large White breed sows with 3 boars of the P-76 French line. The RN genotypes were identified according to Fernandez *et al.* (1990, 1992) on the basis of glycolytic potential measured in biopsy samples and also verified by the technological yield of meat in curing and cooking process, called "Napole yield". Biopsy samples of *Longissimus lumborum* were collected from animals at a weight of 70-80 kilograms according to the Talmant *et al.* (1989) method using PBO/2 apparatus produced by "Biotech" (Slovakia). Glycolytic potential of muscles expressed by lactic acid content in muscle tissue (μmol/g) was calculated according to Monin and Sellier (1985). Fatteners were slaughtered at 100 kilograms liveweight in the same slaughterhouse by electric stunning. The day after slaughter the chilled half carcasses of all animals were subjected to partial dissection according to the method used in Polish Pig Testing Stations for estimation of the carcass quality. In the *Longissimus thoracis* meat quality traits such as pH_1, pH_{24} value, muscle lightness and Water Holding Capacity (WHC) were evaluated. The pH values were recorded using a microcomputer pH meter with a combined glass electrode in muscle homogenates. Muscle lightness was determined using an apparatus Momcolor-D 3098 with white standard. WHC was evaluated according to Grau and Hamm (1952) with the Pohja and Ninivaara (1957) modification. The technological yield of meat in curing and cooking processing (RTN) was evaluated according to the method of Naveau *et al.* (1985).

Moreover, technological yield of meat was defined on cured smoked loin (commercial name - loin "sopocka") which was produced according to the technology utilised in the Polish meat industry.

The comparison between means of genotypes was conducted by t-Student`s test. The effect of the RN- allele was shown as a difference between genotypes RN-rn+ and rn+rn+ value and as a part of this, difference in standard deviation (SD) units (Sellier 1998) and also as percentage part of the average value for the rn+rn+ population.

Results and discussion

Obtained results referring differences of the level glycolytic potential are in agreement with a strong influence of the RN⁻ gene on this trait measured in biopsy samples from live animals (table 1, figure 2). Nowadays it is the best method of identification of pigs with the RN⁻ gene (Le Roy *et al,.* 1994; Monin, 1994; Przybylski, 1998; Przybylski *et al.*, 1998). The investigations also showed the influence of the RN⁻ gene on growth rate. Pigs with genotypes of RN⁻ rn⁺ had around 20 days lower age at slaughter than rn⁺rn⁺ and also smaller backfat thickness (at the level of significance P< 0,07). According to Sellier (1998) carriers of the RN⁻allele also have thinner backfat - 1,3 mm in comparison to rn⁺rn⁺ animals. For other carcass traits no differences between genotypes have been found.

Results of our study according to meat quality showed that the RN⁻ gene in contrast to the halothane sensitivity gene (HAL^n) doesn't influence the rate of after slaughter glycolysis (pH_1) but only the range of changes, measured 24 hours after slaughter (table 1).

Carriers of the RN⁻gene had significantly lower pH ultimate (pH_{24}) and lighter meat colour. This is similar to the results of Monin and Sellier (1985), Naveau (1986) and Lundström *et al.* (1994). According to the above mentioned authors, carriers of the RN⁻ gene have a meat colour close to PSE meat and for distinction it is called "acid meat" or " hampshire meat type". Significantly lower technological yield of meat in curing and cooking process (RTN) and technological yield of smoke-curing loin (TY-SL) was noted for heterozygous RN^-rn^+ than for homozygotes rn^+rn^+ (about 10 and 5% respectively) (table 1). It is probably a result of a low value of pH ultimate (Monin 1994, Przybylski et al., 1994)

Lundström et al. (1994) and Le Roy et al. (1995) obtained similar differences between analogous genotypes for meat quality and its technological yield. In comparison to the HAL gene the decrease mentioned above of technological yield in cooking process was at about 2-3% (Monin 1994, Przybylski et al., 1994).

Table 1. The effect of the RN^- gene on glycolytic potential, growth rate and carcass and meat quality traits in crossbred fatteners.

Traits	$RN^-rn^+ - rn^+rn^+$	$\dfrac{RN^-rn^+ - rn^+rn^+}{rn+rn+} \times 100$	Avg. SE	Significance
Glycolytic potential (μmol/g)	68,68	36,75	19,19	0,05
Age at slaughter (days)	-19,57	-8,69	5,25	0,05
Average back fat thickness (cm)	-0,22	-12,94	0,11	0,07
Loin muscle eye area (cm^2)	0,02	0,05	1,21	NS
Ham without fat and skin (kg)	-0,24	-3,10	0,12	NS
Meat in basic cuts (kg)	-0,16	-0,79	0,46	NS
Meat in carcass (%)	-0,18	-0,34	0,69	NS
pH$_1$	-0,01	-0,16	0,05	NS
pH$_{24}$	-0,17	-3,05	0,03	0,05
Meat lightness	1,14	6,51	0,44	0,05
Water Holding Capacity (cm^2)	0,12	2,35	0,41	NS
RTN	-10,24	-10,68	0,98	0,05
TY-SL	-5,15	-4,44	1,19	0,05

RTN - technological yield of meat in cooking processing; TY-SL - technological yield of smoke-curing loin

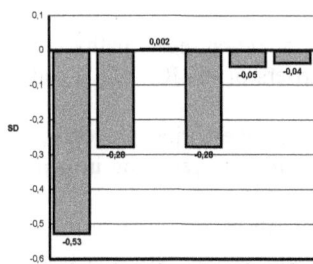

1- age at slaughter, 2- average back fat thickness,
3- loin muscle eye area, 4-ham without fat and skin,
5- meat in basic cuts, 6- meat in carcass

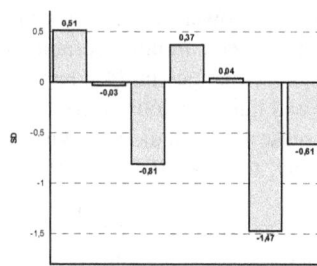

1- glycolytic potential, 2- pH$_1$,
3- pH$_{24}$, 4- meat lightness, 5- WHC, 6- RTN,
7-techn. yield of smoke curing loin (TY-SL)

Figure 1. The effect of the RN^- gene on age at slaughter and carcass traits.

Figure 2. The effect of the RN^- gene on glycolytic potential and meat quality traits.

As a summary of obtained results the effect of the RN¯ gene was calculated as a share of the difference between genotypes in standard deviation (SD) for the population (figures 1 and 2).

The strongest, but unfavourable influence of the described gene, was noted for meat quality traits and their technological yield: RTN (-1,47), pH_{24} (-0,81), TY-SL (-0,61) and PG (0,51). As for the meat lightness the effect of the RN¯ gene was 0,37. Moreover, a relatively high favourable effect was obtained for age when animals gained 100 kilograms liveweight (-0,53), while for back fat thickness the described effect was low (-0,28). The only results analogical to our study were the results of the investigations of Sellier (1998). This author noted, as in our researches, the unfavourable direction of the influence of the RN¯ gene on meat quality and technological yield (RTN), but the effect was stronger. Similar results concerned slaughtering performance traits such as average back fat thickness and lean meat content in carcass.

In this paper interesting results enriching our knowledge in the domain of influence of the described gene on other important traits were also obtained. Results mentioned above concern the effect of the RN¯ gene on the technological yield of meat in the process production of smoked loin (TY-SL) and growth rate measured as age of 100 kilogram liveweight gain (-0,61 and -0,53 respectively).

Conclusions

1. Investigations confirm that the RN¯ allele increases the glycolytic potential value at about 37%, measured in biopsy samples of *Longissimus lumborum* muscle.
2. The obtained results show that the RN¯ allele influences the growth rate and back fat thickness. Pigs with genotypes RN¯rn⁺ had around 20 days lower age at slaughter than rn⁺rn⁺ and also smaller average back fat thickness.
3. The results showed the unfavourable effect of the RN¯ allele on meat quality traits, which define range of changes in 24 hours *post mortem*. Carriers of the RN¯ allele had lower pH ultimate and lighter colour of meat.
4. Investigations confirm that the RN¯ allele increases the glycolytic potential value at about 37%, measured in biopsy samples of *Longissimus lumborum* muscle.
5. The obtained results show that the RN¯ allele influences the growth rate and back fat thickness. Pigs with genotypes RN¯rn⁺ had around 20 days lower age at slaughter than rn⁺rn⁺ and also lower average back fat thickness.
6. The results showed an unfavourable effect of the RN¯ allele on meat quality traits, defined by a range of changes in 24 hours *post mortem*. Carriers of the RN¯ allele had lower pH ultimate and lighter colour of meat.
7. The decrease of technological yield in cooking and smoke-curing processes was noted respectively at about 10% and 5% for pigs with the RN¯rn⁺ genotype in comparison to rn⁺rn⁺.
8. Statistically significant differences between genotypes RN¯rn⁺ rn⁺rn⁺ in phenotypic values of traits expressed in SD units reflecting strength and direction of influence of this gene on analysed traits was as follows: -0,53 for age of slaughter, -0,28 for average back fat thickness, 0,51 for glycolytic potential, -0,81 for pH_{24}, 0,37 for meat lightness, -1,47 for RTN and -0,61 for technological yield of smoke-curing loin processing.

References

Fernandez, X., J. Naveau, A. Talmant & G. Monin, 1990. Distribution du potential glycolytique dans une population porcine et relation avec le rendement Napole. Journées de la Recherche Porcine en France 22: 97-100

Fernandez, X., E. Tornberg, J. Naveau, A. Talmant & G. Monin, 1992. Bimodal distribution of the muscle glycolytic potential in French and Swedish populations of Hampshire crossbred pigs. Journal of the Science of Food and Agriculture 59: 307-311

Grau, R. & R. Hamm, 1952. Eine einfache Methode zur Bestimmung Wasserbindung in Fleisch. Fleischwirtschaft 4: 295-297

Le Roy, P., J.C. Caritez, Y. Billon, J.M. Elsen, A. Talmant, P. Vernin, H. Lagant, K. Larzul, G. Monin & P. Sellier, 1995. Étude de l'effet du locus rn sur les caractères de croissance et de carcasse. Journées de la Recherche Porcine en France 27: 165-170

Le Roy, P., W. Przybylski, T. Burlot, C. Bazin, H. Lagant & G. Monin, 1994. Etude des relations entre le potentiel glycolytique du muscle et les caracteres de production dand les lignees Laconie et Penshire. Journées de la Recherche Porcine en France 26: 311-314

Lundström, K., A. Andersson, S. Maerz, I. Hansson, 1994. Effect of the RN- gene on meat quality and lean meat content in crossbred pigs with hampshire as terminal sire. 40th ICoMST, The Hague, Netherlands, S-IV.A.07

Milan, D., P. Le Roy, N. Woloszyn, J.C. Caritez, J.M. Elsen, P. Sellier & J. Gellin, 1995. The RN locus for meat quality maps to pig chromosome 15. Genetics, Selection, Evolution 27: 195-199

Monin, G., 1994. Effects of the RN gene on pig meat quality. IInd Int. Conf. "The influence of genetic and non genetic traits on carcass and meat quality", Siedlce, 7-8 Nov. 1994: 37-48

Monin, G. & P. Sellier, 1985. Pork of low technological quality with a normal rate of muscle pH fall in the immediate post-mortem period: The case of the hampshire breed. Meat Science 13: 49-63

Naveau, J., 1986. Contribution à l'étude du déterminisme génétique de la qualité de viande porcine. Héritabilité du Rendement Technologique Napole. Journées de la Recherche Porcine en France 18: 265-276

Naveau, J., P. Pommeret & P. Lechaux, 1985. Proposition d'une méthode de mesure du rendement technologique : la "méthode Napole". Techni-Porc 8: 7-13

Pohja, N.S. & F.P. Ninivaara, 1957. Die Bestimmung der Wasserbindung des Fleisches mittels der Konstandrückmethods. Fleischwirtschaft 9: 193-195

Przybylski, W., 1998. A method for detection of pigs with predisposition to producing of PSE and acid meat. Polish Journal of Food and Nutrition Sciences, Vol. 7/48, No. 4 (S): 246-250

Przybylski, W., M. Koćwin-Podsiadła, S. Kaczorek, E. Krzęcio, J. Kurył, 1994. Effect of the HALn gene in heterozygous pigs on the carcass traits, fresh meat quality and its technological yield. IInd International Conference: "Influence of genetic and non genetic traits on carcass and meat quality". Siedlce: 176-183

Przybylski, W., M. Koćwin-Podsiadła, J. Kurył & G. Monin, 1996. Meat quality in two genetic groups of pigs with RYR1 and RN- genes. 42nd ICoMST, Lillehammer, Norway, H.1: 294-295

Przybylski, W., M. Koćwin-Podsiadła, E. Krzęcio & S. Kaczorek, 1998. The frequency of RN- gene in different group of pigs. Polish Journal of Food and Nutrition Sciences, Vol. 7/48, No. 4 (S): 234-240

Sellier, P., 1998. Genetics of meat and carcass traits. In: The genetics of the pig. (ed. M.F. Rothschild & A. Ruvinsky), CAB International, Wallingford, Oxon: 463-510

Talmant, A., X. Fernandez, P. Sellier, & G. Monin, 1989. Glycolytic potential in Longissimus dorsi muscle of Large White pigs, as measured after in vivo sampling. 35th ICoMST, Copenhagen, Denmark, 6.33: 1129-1131

Breed and slaughter weight effects on meat quality traits in hal- pig populations

X. Puigvert [1,2], J. Tibau [*2], J. Soler [2], M. Gispert [2] & A. Diestre [2]

[1] EPS – Universitat de Girona. Avda L. Santaló s/n, 17003 Girona, Catalunya, Spain
[2] IRTA – CCP – CTC. Veïnat de Sies s/n, Monells, 17121 Girona, Catalunya, Spain

Summary

Halothane negative pig populations are the optimal choice to produce female lines in modern pig production schemes. The halothane gene has a detrimental effect on meat quality traits in heterozygous populations, thus it is important to characterise the maternal lines free of the halothane gene. This contribution analysed several meat characteristics on 110 entire males slaughtered at 90 and 110 kilograms liveweight. Animals were tested to be non-carriers of the halothane gene, were of Large White, Landrace and Duroc breeds and belonged to the National Pig Breeders Association.

Duroc animals reveal higher pH values (at 45 minutes and 24 hours *post mortem*). *Longissimus toracis* muscles of Landrace and Large White animals were paler (subjective colour scale) compared with Duroc ones. The main difference between Duroc and white breeds is their higher intramuscular fat content. Slaughter weight doesn't affect meat quality characteristics. Breed x Slaughter weight interactions were not significant in most meat traits analysed.

Keywords: Pigs, Halothane genotype, Breed, Slaughter weight, Meat quality

Introduction

The Spanish pig meat industry demands heavier carcasses, and meat quality is progressively better appreciated by the consumer. Several specialised sire lines are used to produce lean carcasses for fresh consumption or to supply the dry cured ham industry. Alternative crosses to produce hybrid sows are currently based on breeds with recognised reproductive performances and free of the halothane gene.

The knowledge of meat quality traits of these maternal lines is important to establish a long term crossbreeding strategy, because the values of heterosis effects for body composition and meat quality traits are close to zero (Sellier, 1976).

This work aims to analyse the breed effect on meat quality traits in the Large White, Landrace and Duroc breeds, slaughtered at two target liveweights (90 kilograms and 110 kilograms), taking into account the possible interactions between the main effects.

Material and Methods

This study was undertaken with 110 piglets from the Large White (LW), Landrace (LR) and Duroc (DU) breeds provided by 18 nucleus herds belonging to the Spanish National Association of Pig Breeders (ANPS) and born within one week of each other. The halothane genotype was determined using the molecular analysis of the cDNA from blood samples

171

(Fujii *et al.,* 1991; Sanchez-Bonastre *et al.*, 1993). In total, 40 Large White, 34 Landrace and 36 Duroc entire males were involved in this study.

The animals were reared in two batches at the Pig Testing Centre (IRTA) and were fed *ad libitum* with the same diet until they reached the allocated slaughter weight. Half of the animals of each breed were randomly slaughtered at each target weight (90 and 110 kilograms). The animals were slaughtered at the Meat Technology Centre (IRTA), after a standardised pre-slaughter treatment (3 km of transport, 12 hours in lairage with access to water and electrically stunned -350 V at 50 Hz-).

The left side of the carcass was used to assess meat quality. The following variables of meat quality were recorded:

- Muscle pH, measured at the last rib level of *Longissimus toracis* (LT) and *semimembranosus* muscle (SM) at 45 minutes (pH45) and 24 hours (pH24) *post mortem*, using a portable pHmeter (Scharlau, Hl-9025) equipped with a xerolyt electrode.
- Electrical conductivity (Pork Quality Meater (QM), INTEK, Aichach, Germany), measured at the last riblevel of LT and SM at 45 minutes (QM45) and 24 hours (QM24) *post mortem.*
- Muscle colour at 24 hours *post mortem* on the exposed cut surface of the LT, at the last rib using a Minolta Chromameter CR-200 and determining lightness (L*) and chromaticity dimensions (a* and b* values) and visual judgement of colour using the 6-point scale described by Nakai *et al.* (1975).
- Intramuscular fat (IMF) was determined in a sample of 200 gr of LT taken at the last rib and free of connective tissue, using the Near Infrared Transmittance system (Infratec 1265 Meat Analyzer).

All the analyses were performed using the General Linear Model program of the SAS statistical package (SAS, 1988). Fixed effects of breed (B), slaughter weight (W), batch and their interactions were included in the analysis by least-square means. Deviation of individual weight from the mean weight of 90 kilograms or 110 kilograms, was included as a covariate. The following general model was used:

$$Y_{ijk} = \mu + B_i + W_j + (B \times W)_{ij} + b_k(\mu_k - W_{ijk}) + e_{ijk}$$

where Y_{ijkl} = dependent variable, μ = overall mean of the trait, B_i = breed effect (LW and LR), W_j = slaughter weight effect (90 kg and 110 kg), b_k = intragroup slaughter weight regression coefficient, e_{ijk} = residual random term. No batch effect was observed in the experience.

Results and Discussion

Results in table 1 indicate good meat quality characteristics in all three breeds. Average figures fit the thresholds (pH45' < 5.8 and L > 56) for non-PSE (Pale, Soft and Exudative) meat.

Large White and Landrace animals showed lower LT pH values than the Duroc breed. Visual appraisal colour was paler in these two breeds compared with the Duroc.

The most interesting factor associated with the Duroc breed is the effect of intramuscular fatness. The intramuscular fat concentration in Duroc LT muscle (1.66 %) is significantly higher than in Large White and Landrace animals. The figures are based on entire male animals, but higher values (20-30 % more) could be obtained in castrate animals. Marbling has a positive effect on taste characteristic and tenderness (Barton-Gade & Bejerholm, 1985)

Table 1. Least-square means of meat quality characteristics with the significant interactions (P < .05).

	Breed			Slaughter Weight		Large White		Landrace		Duroc	
	LW	LD	DU	90 kg	110 kg	90 kg	110 kg	90 kg	110 kg	90 kg	110 kg
pH45LD	6.26[a]	6.30[a,b]	6.41[b]	6.31	6.33	--	--	--	--	--	--
pH45SM	6.26	6.28	6.32	6.25	6.32	--	--	--	--	--	--
pHuLD	5.66[a]	5.65[a]	5.76[b]	5.69	5.69	5.63[a,b]	5.68[a,b]	5.70[a,b]	5.59[a]	5.73[b,c]	5.80[c]
pHuSM	5.63	5.63	5.64	5.64	5.62	5.63[a,b,c]	5.63[a,b,c]	5.68[a,c]	5.57[b]	5.62[b,c]	5.65[a,c]
QM45LD	4.50	4.62	4.72	4.73	4.50	--	--	--	--	--	--
QM45SM	4.49	4.57	4.69	4.54	4.62	--	--	--	--	--	--
QmuLD	3.94	4.24	3.88	4.13	3.91	3.71[a]	4.18[ab]	4.91[b]	3.56[a]	3.77[a]	3.98[a,b]
QmuSM	5.13	5.26	5.67	5.04	5.67	--	--	--	--	--	--
Colour	2.86[a,b]	2.79[a]	3.06[b]	2.90	2.91	--	--	--	--	--	--
L*	49.80	51.55	49.22	50.98	49.40	--	--	--	--	--	--
a*	7.33	8.05	7.96	7.34[a]	8.23[b]	--	--	--	--	--	--
b*	4.61[a]	5.28[b]	4.85[a,b]	4.68	5.14	--	--	--	--	--	--
IMF	0.97[a]	0.87[a]	1.66[b]	1.10	1.23	--	--	--	--	--	--

[a,b] Different letters within main effects indicate significant differences (P < .05).
Longissimus muscle (LT) and Muscle semimembranosus (SM).
QM: muscle electrical conductivity (µs).
pH45, QM45; at 45 min post mortem.
pHu, Qmu; at 24 h post mortem.
Colour: Japanese subjective colour scale (1;pale – 6; dark)
IMF: Intramuscular fat content (%).

In our study, slaughter weight appears to exert a small effect on meat quality characteristics. García-Macías et al., (1996), Candek-Potokar et al., (1996) and Leach et al., (1996) also found no effect of slaughter weight on the majority of the meat quality variables. Only a significant effect of carcass weight on chromameter a* value (p < 0.05) was found. With increasing weight, muscle became redder in colour, probably due to an increase in muscle pigment concentration. Similar results were obtained by García-Macías et al., (1996). No significant increase of intramuscular fat was found with increasing slaughter weight.

Only breed*slaughter weight interactions are found in pHu values in LD and SM muscles and in QMu values in LD muscle.

Acknowledgements

This work was supported by the Insitituto Nacional de Investigaciones Agrarias (INIA – SC 95-048). Special thanks to ANPROGAPOR, ACPS, and ANPS pig producer associations for providing the animals.

References

Barton-Gade, P. & C. Bejerholm, 1985. Eating quality of pork. Pig Farming Supplement Dec. 1985: 56-57

Candek-Potokar, M., B. Zlender & M. Bonneau, 1996. Quality parameters of pig Longissimus Dorsi muscle as affected by slaughter weight and breed. Proc. of the 42nd International Congress of Meat Science and Technology: 306-307

Fujii, J., K. Otsu, F. Zorzato, S. de Leon, V. K. Khanna, J. E. Weiler, P. J. O'Brien & D. H. MacLennan, 1991. Identification of a mutation on porcine ryanodine receptor associated with malignant hyperthermia. Science 253: 448-451

García-Macías, J. A., M. Gispert, M. A. Oliver, A. Diestre, P. Alonso, A. Muñoz-Luna, K. Siggens & D. Cuthbert-Heavens, 1996. The effects of cross, slaughter weight and halothane genotype on leanness and meat and fat quality in pig carcasses. Animal Science 63: 487-496

Leach, L. M., M. Ellis, D. S. Sutton, F. K. McKeith & E. R. Wilson. 1996. The growth performance, carcass characteristics, and meat quality of halothane carrier and negative pigs. Journal of Animal Science 74: 934-943

Sanchez-Bonastre, A., J. M. Folch-Albareda & A. Coll-Cerdà, 1993. Optimización del análisis molecular para la detección del síndrome de estrés porcino. V jornadas sobre Producción Animal, Zaragoza: 257-259

SAS, 1988. SAS/STAT user's guide: release 6.03 edition. Statistical Analysis Systems Institute Inc., Cary, NC

Sellier, P, 1976. The basis of crossbreeding in pigs; a review. Livestock Production Science 3: 203-226.

Genotypic and allelic frequencies of the RYRI locus in the Manchado de Jabugo pig breed

A.M. Ramos[1], J.V. Delgado[2], T. Rangel-Figueiredo[1], C. Barba[2], J. Matos[3] & M. Cumbreras[4]

1. Div. Fisiologia Animal Departamento Zootecnia Univerdidade Trás-os-Montes e Alto Douro Apartado 202 5001 Vila Real Codex Portugal.
2. Departamento de Genética. Universidad de Córdoba. Av. Medina Azahara, 9 14005 Córdoba. Spain.
3. INETI/IBQTA/DB/BQ II Estrada Paço Lumiar 1699 Lisboa Codex Portugal
4. Servicio de Ganadería. Diputación Provincial de Huelva. Apdo. 26. Huelva . Spain.

Summary

Manchado de Jabugo is a breed related to the Iberian Pig branch, even though the origin of this breed includes, together with the Iberian Pig, other breeds from England and Germany, from the first half of the present century. This study is part of an ambitious investigation of the implication of the variation of the RYR1 locus in the breeds of the Iberian Peninsula, with a view to determine the grade of expected effects of the porcine stress syndrome, on the native pig populations of Spain and Portugal. Manchado de Jabugo, because of its origin could be an entry point for high frequencies of the recessive allele from foreign breeds to the Iberian representatives of the Mediterranean pig breeds, traditionally admitted to have a low frequency for PSS. In this study we have employed the PCR-RFLP method for genotyping this locus in a sample of animals of both sexes belonging to the breed. The steps of the technique were as follows: amplification of the locus fragment and digestion with the restriction enzyme *Hha*I. We are presenting the genotypic and allelic frequencies of the variants of this locus in both sexes, resulting in a null presentation of the **n** allele, in neither the homozygous nor the heterozygous form, so we can reach important conclusions about the relationships among the diverse Iberian breed populations of Spain and Portugal.

Keywords: Conservation, Characterisation, DNA markers, PSS

Introduction

Sometimes, the native breeds are carriers of interesting qualities that could lonely justify any actions done for their conservation. For this purpose our two teams belonging to the University of Cordoba (Spain) and the University of Trás-os-Montes e Alto Douro (Portugal) are involved in a collaboration project with a view of evaluating the genotypic frequency of the RYR1 locus (ryanodine receptor gene) in all native pig breeds of the Iberian Peninsula. In this paper we are showing our results obtained in the study of the Manchado de Jabugo breed, a small Spanish pig population (less than 200 animals) which is near extinction. The Manchado de Jabugo is a population belonging to the Mediterranean pig branch, strongly linked to the Iberian Pig (Rodero *et al.*, 1994). This breed is located in southwestern Spain and was described for the first time by Mateos (1967) and is now internationally recognised (Porter, 1993; DAD-IS, 1996). These animals were extensively reared in the first half of the present century, but they suffered under the pressure of the international breeds (Large White,

Piétrain, and Landrace) and some variants of the Iberian Pig. At present the last 200 animals are part of a specific conservation program.

The RYR1 locus affects several quantitative and qualitative meat traits (Goodwin et al,. 1994; Louis et al., 1994). Its recessive allele (n) is responsible for the presence of the porcine stress syndrome (PSS), when it appears in the homozygous form. Sometimes this syndrome leads to the appearance of the illness malignant hyperthermia (MH). It is very important to know the frequency of this recessive allele in a determined breed because a low presence constitutes an extremely good characteristic in the definition of such a population, and it could justify its conservation when we are working with an endangered breed.

The Manchado de Jabugo is mainly exploited for the quality of its meat, destined for the preparation of manufactured products (loin, ham, etc). In this type of breed it is specially relevant to know on one hand the recessive gene frequency, and on the other hand to identify the possible carriers of the recessive allele for selection purposes.

Material and Methods

We have used in the present study 46 animals belonging to the Manchado de Jabugo pig breed, 24 females and 22 males, distributed in the last two herds where these animals are bred in purity. It constitutes 30 % of the total of the breed's population. DNA was extracted from hair samples, according to the method described by Kawasaki (1990).

PCR was developed in 25 µl reaction using specific primers for the amplification of a 134 bp fragment of the RYR gene, which contained the C to T transition at nucleotide 1843. The primers sequences were: RYR1 (5'-GTGCTGGATG-TCCTGTGTTCCCT-3') and RYR2 (5'-CTGGTGACATAGTTGAT-GAGGTTTG-3') (Brenig & Brem, 1992; Russo et al., 1993). The reaction mixture contained: 10× PCR buffer, 200 µM of each dNTP, 1.5 mM $MgCl_2$, approximately 200 ng of template DNA, 10 pmol of each primer and 1 U of Taq polymerase (Promega, Madison, USA). The samples were submitted to 34 cycles of amplification, each cycle consisting of: 40 s at 95°C, 30 s at 60°C and 30 s at 72°C. Finally, the PCR products were incubated overnight at 37°C with 10 U of the enzyme HhaI. The digested fragments were separated by electrophoresis in 4% agarose gels and observed by ethidium bromide staining. The Biocapture software (Vilber Lourmat, Marne La Valée, France) was employed to analyse the DNA profiles.

Results and Discussion

All the animals sampled in the present work have shown a homozygous NN genotype, which corresponds to animals resistant to PSS. If we take into account the extreme representativeness of the sample (30 % of the total breed population) we can consider this population as free of the recessive n allele. It constitutes an important characteristic of this breed overall when its animals are destined to the production of high quality meat for transformed products.

Our results are in accordance with those published by Matassino et al. (1998) concerning three Italian native pig breeds. Our breeds, like the Italian ones, belong to the Mediterranean branch (Porter, 1993). Therefore we can assume a low presence of the recessive allele in this important evolutive group of animals. However, Ramos (personal communication) has found similar frequencies of the recessive allele in the Portuguese Alentejano breed as in other pig breeds belonging to other evolutive branches, such as the Bísaro (Ramos, personal communication), Yorkshire (Houde et al., 1993) and Italian Landrace breeds (Russo et al., 1996).

From our point of view we can consider that from an evolutive perspective, the mutation on the 1843 nucleotide of the RYR locus that produces the **n** recessive allele could happen in populations belonging to the Celtic northern pig branch. It could explain the demonstrated null presence of this allele in four Mediterranean pig breeds. The arrival of this allele to breeds such as the Alentejano breed could be explained by the contamination of these breeds with some imported breeds specialising in the precocity of growth.

Currently we are developing similar studies on other Spanish native pig breeds, particularly all the varieties of the Iberian pig breed, in order to go deeply into the behaviour of the **n** recessive allele in the Iberian peninsula, where we have representation of other native breeds belonging to the Mediterranean Type (Iberico, Chato Murciano and Negro Mallorquin), the Celtic type (Celta Gallego), and also the ancient African branch (Negro Canario). This study could give us a good perspective on the genesis and evolution of the PSS disgenesic character, associated with the production of pale, soft and exudative meat (PSE) responsible for great economic damage to the pig industry.

At present the traditional extensive farming of native breeds in Spain is demanding faster-growing animals. The possible introduction in these breeds of the **n** allele from foreign breeds which are presently contaminating these indigenous populations, is dangerous because MH is traditionally linked to a higher and faster muscle development. It implies a rapid extension of the recessive allele in those populations supported by these selection criteria in the present breeding programs.

References

Brenig, B. & G. Brem, 1992. Molecular cloning and analysis of the porcine «halothane» gene. Arch. Tierz. Dummerstorf, 35: 129-135

DAD-IS, 1997. Domestical Animal Diversity Information System. http://www.fao.org/dad-is/

Goodwin, R.N., L.L. Christian, C.F. Louis, P.J. BERGER, K.J. Prusa & D.C. Beitz, 1994. Effect of the HAL gene on pork carcass composition and eating quality traits. J. Anim. Sci. 72: S 958

Houde, A., S.A. Pommier & R. Roy, 1993. Detection of the ryanodine receptor mutation associated with malignant hyperthermia in purebred swine populations. J. Anim. Sci.: 71: 1414-1418

Kawasaki, E., 1990. Sample preparation from blood cells and other fluids. In: PCR protocol: A guide to methods and applications. Academic Press Inc., New York, 146-152

Louis, C. F., J.R. Mickelson & W.E. Rempel, 1994. The effect of skeletal muscle ryanodine receptor (ryr1) genotype on swine performance and carcass traits. J. Anim. Sci.: 72, S 955

Matassino, D., R. Davoli, M. Occidente, J. Milc, G. Caiola, M. Rocco, & V. Russo, 1998. Identificazione del genotipo per la sensibilità all'alotano in alcuni tipi genetici suini autoctoni. IV Simpósio Internacional do Porco Meditterânico, Évora, Portugal, 26-28 Novembro

Mateos, B., 1967. Origen y estandar de la raza Manchada de Jabugo en sus dos variedades. Arch. Zoot. 16: 317-340

Porter, V., 1993. Pig, a handbook to the Breeds of the World. The banks, Mountfield, Near Robersbrigde, East Sussex

Ramos, M.A., 1999. Personal communication

Rodero, E., J.V. Delgado, M.E. Camacho & A. Rodero, 1994. Conservación de razas andaluzas en peligro de extinción. Ediciones Junta de Andalucía. Sevilla

Russo, V., S. Dall'Olio, R. Davoli, M.B. Coscelli & D. Bigi, 1996. Studio del locus alotano nelle raze suine allevate in Italia mediante test PCR. Zoot. Nutr. Anim. 22: 33-38

Russo, V., R. Davoli, J. Tagliavini, S. Dall'olio, D. Bigi, E. Costosi, M.B. Coscelli & L. Fontanesi, 1993. Identificazione del genotipo dei suini per la sensibilità all'alotano a livello di DNA mediante PCR. Zoot. Nutr. Anim., 19: 89-93

Acknowledgements

This work is enclosed in the EU project GENRES012. We would like to thank the Diputacion Provincial de Huelva for the facilitation of the blood sample collection.

Variation of meat quality in pig breeds

Intramuscular fat content in some native German pig breeds

U. Baulain[1], P. Köhler[1], E. Kallweit[1] & W. Brade[2]

[1]*Institut für Tierzucht und Tierverhalten (FAL), D-31535 Neustadt-Mariensee, Germany*
[2]*Landwirtschaftskammer Hannover, D-30159 Hannover, Germany*

Summary

Sows and castrates of some native German pig breeds, the Angeln Saddleback, Bentheimer Black Pied and Swabian Hall Saddleback and their crosses with Piétrain were fattened at a testing station and compared to Large White x German Landrace, Duroc x German Landrace and to purebred Piétrain. Due to the standard feeding scheme at the testing station sows were fed *ad libitum* but castrates restricted. Fattening, carcass and meat quality data of 399 pigs were recorded. Percentage of lean was determined by FOM. Intramuscular fat content (IMF) of *M. longissimus dorsi* was predicted by means of near infrared transmission spectroscopy. The highest IMF was found in Swabian Hall Saddleback, Duroc x German Landrace and Angeln Saddleback castrates with 1.64%, 1.58% and 1.42%, respectively. IMF of less than 0.70% was observed in Piétrain sows and castrates as well as in female Swabian Hall Saddleback crosses. Thus IMF in native breeds was less than expected. The lowest pH values (45 minutes) of *M. longissimus dorsi* were found in purebred Piétrain and Swabian Hall Saddleback crosses with 5.52 and 5.87, respectively. An overall correlation of r = 0.46 between IMF and pH was observed. The highest percentage of lean was noticed in female Piétrain and Swabian Hall Saddleback crosses and in Piétrain castrates with 61.9%, 57.4% and 57.0%, respectively. Portion of lean was less than 46.5% in Bentheimer Black Pied and Swabian Hall Saddleback castrates. An overall correlation of r = -0.58 between percentage of lean and IMF was determined.

Keywords: Meat quality, Pigs, Intramuscular fat, Near infrared spectroscopy

Introduction

During the last ten years more emphasis has been given to studying the intramuscular fat content in pigs (Brandt, 1996; Schwörer *et al.*, 1996). One reason is the consumer's demand for a better meat quality. Meat with a higher intramuscular fat is expected to have a better sensoric quality. As shown in previous studies (Kallweit *et al.*, 1996) intramuscular fat is relatively low in most German pig breeds, with the exception of Duroc and their crossbreds. The aim of this investigation was to compare carcass and meat characteristics of some native German pig breeds and their crosses with Piétrain to breeds currently used in production. The old breeds are mostly kept as gene reserves and are expected to have a very good meat quality. A description of these breeds is given by Sambraus (1986).

Material and methods

Sows and male castrates of some native German pig breeds, Angeln Saddleback (AS), Bentheimer Black Pied (Be) and Swabian Hall Saddleback (SH) and their crosses with Piétrain were fattened at a testing station of the Chamber of Agriculture Hannover and compared with Large White x German Landrace (DE*DL), Duroc x German Landrace (Du*DL) and purebred Piétrain (Pi). Due to the standard feeding scheme at the testing station sows were fed *ad libitum* but castrates restricted. Therefore the effect of sex could not be estimated separately. Fattening, carcass and meat quality data of 399 pigs were registered according to the regular testing procedure. All pigs were slaughtered in the experimental slaughterhouse at Mariensee from February to April 1996. Average slaughter weight was 84.8 kilograms. Percentage of lean was determined by Fat-O-Meter (FOM). The intramuscular fat content (IMF) of *M. longissimus dorsi* at 13/14 thoracic vertebrae was predicted by means of near infrared transmission spectroscopy (NIT). All animals were included in the analysis, even if IMF was less than 0.5%, which is the lower limit of the prediction equation used by NIT.

Results

Due to the instructions of the station testing, the pigs should have a slaughter weight of 85 kilograms. From table 1 it is evident that the deviations from this target weight were not very large.

In table 1 it is also demonstrated that purebred castrates of the native breeds Swabian Hall Saddleback, Angeln Saddleback and Bentheimer Black Pied German breeds had the lowest percentages of lean with clearly less than 50% whereas purebred female Piétrain pigs yielded more than 60% lean meat.

Table 1. *Slaughter weight and percentage of lean predicted by FOM in castrated male and female pigs of different breeds (means and standard deviations).*

breed	m/f	slaughter weight (kg)				lean (%)			
		sex = m		sex = f		sex = m		sex = f	
	N	$\bar{x}$	s_x	$\bar{x}$	s_x	$\bar{x}$	s_x	$\bar{x}$	s_x
AS	15/12	82.5	2.4	84.1	3.0	48.2	3.6	48.2	2.7
Be	19/21	84.1	2.3	85.0	2.2	46.4	3.3	50.3	2.9
SH	36/0	82.5	1.7	--	--	46.3	4.1	--	--
Du * DL	36/0	83.4	2.5	--	--	54.3	2.4	--	--
DE * DL	101/0	84.6	2.8	--	--	53.2	2.9	--	--
AS * Pi	18/14	85.1	2.8	87.1	4.0	49.7	3.4	53.2	3.8
SH * Pi	33/38	85.2	2.5	86.7	2.2	54.1	2.6	57.4	2.3
Pi	15/41	87.2	3.2	86.2	3.3	57.0	1.9	61.9	1.7

In table 2 it is shown that castrates of the native breeds and also of Duroc x German Landrace had the highest IMF of about 1.5%. A very low IMF of less than 0.7% was observed in Piétrain sows and castrates but also in female Swabian Hall Saddleback crosses.

The mean values for lean and IMF show the antagonism between these traits as well as the correlations presented in table 4, which were significant within each breed and over all animals. Only pigs carrying the Duroc gene seem to be an exception. Although these pigs had a high IMF of 1.58% the percentage of lean was 54.3%. Compared to the other castrates only the purebred Piétrain pigs had a higher percentage of lean than the Durocs.

Table 2. Backfat thickness (cm) and intramuscular fat content (IMF) in M. longissimus dorsi of castrated male and female pigs of different breeds (means and standard deviations).

| breed | m/f | backfat thickness (cm) | | | | IMF (%) | | | |
| | | sex = m | | sex = f | | sex = m | | sex = f | |
	N	x	s_x	x	s_x	x	s_x	x	s_x
AS	15/12	3.06	0.33	3.11	0.34	1.42	0.69	1.05	0.40
Be	19/21	3.14	0.43	2.86	0.31	1.36	0.68	0.95	0.49
SH	36/0	3.13	0.40	--	--	1.64	0.77	--	--
Du * DL	36/0	2.36	0.37	--	--	1.58	0.74	--	--
DE * DL	101/0	2.51	0.33	--	--	0.85	0.43	--	--
AS * Pi	18/14	3.26	0.52	2.79	0.47	1.03	0.33	0.76	0.37
SH * Pi	33/38	2.66	0.32	2.24	0.28	0.90	0.37	0.59	0.21
Pi	15/41	2.33	0.23	1.81	0.27	0.69	0.25	0.41	0.20

Table 3 indicates that the meat quality characterised by the lowest values for pH_{45min} and the highest values for LF_{24h} (electrical conductivity) was not sufficient in purebred Piétrain or in Swabian Hall Saddleback crosses. For all other breeds pH values of more than 6.0 were observed.

Table 3. pH_{45min} and electrical conductivity (LF_{24h}) of different breeds (means and standard deviations).

| breed | N | pH_{45min} | | LF_{24h} | |
		x	s_x	x	s_x
AS	27	6.18	0.28	3.89	2.00
Be	40	6.16	0.32	4.07	1.96
SH	36	6.25	0.32	3.54	1.26
Du * DL	36	6.33	0.22	2.93	0.75
DE * DL	101	6.04	0.38	4.69	2.10
AS * Pi	32	6.15	0.38	5.08	2.20
SH * Pi	71	5.87	0.25	6.52	1.68
Pi	56	5.52	0.21	8.11	1.64

Besides the negative correlation mentioned above between IMF and percentage of lean, table 4 shows the positive correlation between backfat thickness and IMF. In regard to the meat quality traits pH and electrical conductivity, a clear overall relationship to IMF was found. But within breed the correlations seemed to be different.

Table 4. Correlations between intramuscular fat content and different carcass and meat quality traits (correlations within breed and overall).

breed	N	lean	backfat	pH_{45min}	LF_{24h}
AS	27	-0.50	0.18	0.44	-0.49
Be	40	-0.45	0.27	0.36	-0.27
SH	36	-0.59	0.27	0.11	0.15
Du * DL	36	-0.33	0.45	0.14	-0.37
DE * DL	101	-0.41	0.15	0.30	-0.23
AS * Pi	32	-0.53	0.33	0.15	-0.11
SH * Pi	71	-0.57	0.52	0.07	-0.08
Pi	56	-0.52	0.48	0.29	0.00
overall	399	-0.58	0.43	0.46	-0.44

Conclusions

As expected, castrates of the native breeds Swabian Hall Saddleback, Angeln Saddleback and Bentheimer Black Pied and also Duroc x German Landrace had the highest IMF but clearly less than 2% which is supposed to be a sufficient value. Logically, the lowest IMF of less than 0.7% was observed in Piétrain pigs. Compared to purebreds, crossbreds of native AS and SH with Piétrain had a clearly lower IMF.

Percentage of lean was significantly lower in the native breeds. The relatively high negative correlation between percentage of lean and IMF demonstrates the antagonism between these traits, which has to be considered in the case of selection on IMF.

Measures of pH and electrical conductivity show that meat quality in Piétrain and Swabian Hall Saddleback crosses was not sufficient. The negative correlation between IMF and pH indicates that IMF serves as an energy source in muscle metabolism and has a positive influence on p. m. processes.

Meat quality in purebred native breeds does not seem to be superior to breeds currently used in pig production and the lean content of these breeds must be considered as too low.

References

Brandt, H., 1996. Möglichkeiten der Zucht auf höheren intramuskulären Fettgehalt unter deutschen Marktverhältnissen. IMF-Kolloquium der Thüringer Landesanstalt für Landwirtschaft, Wilhelmsthal, 22.-23.10.1996: V17.

Kallweit, E., M. Henning, P. Köhler & U. Baulain, 1996. Intramuskulärer Fettgehalt bei verschiedenen Schweinerassen. IMF-Kolloquium der Thüringer Landesanstalt für Landwirtschaft, Wilhelmsthal, 22.-23.10.1996: V18.

Sambraus, H. H., 1986. Atlas der Nutztierrassen: 180 Rassen in Wort und Bild. Eugen Ulmer, Stuttgart, Germany, 256 pp.

Schwörer, D., A. Hofer, D. Lorenz & A. Rebsamen, 1996. Selektion auf intramuskuläres Fett in der Schweizerischen Schweinezucht. IMF-Kolloquium der Thüringer Landesanstalt für Landwirtschaft, Wilhelmsthal, 22.-23.10.1996: V19.

Fatty acid composition and cholesterol content of the fat of pigs of various genotypes

J. Csapó[1], F. Húsvéth[2], Zs. Csapó-Kiss[1], P. Horn[1], Z. Házas[1] & É. Varga-Visi[1]

[1]*Pannon University of Agriculture (PATE) Faculty of Animal Science, Kaposvár, Hungary*
[2]*PATE Georgikon Faculty of Agricultural Science, Keszthely, Hungary*
[3]*HUNGAPIG Co. Ltd., Herceghalom, Hungary*

Summary

The authors determined the fatty acid composition and the fat cholesterol content of the fat of Mangalica, Hungarian Large White x Hungarian Landrace and Mangalica x Duroc pigs. It was established that no significant difference among the three genotypes could be detected with respect to saturated, unsaturated, or the essential fatty acids, nor in regard to cholesterol content. The findings of these investigations indicate that in the three pig genotypes studied, fat cholesterol content varies between 71 and 109 mg/100 g. Attention is also drawn to the high oleic acid content (relative %age 43.57-44.81) and linoleic acid content (relative %age 10.63-11.47) of pig fat.

Keywords: Fatty acids, Cholesterol content, Fat, Pigs, Various genotypes

Introduction

The fatty acid composition of the fat content of foodstuffs is of extremely great importance with respect to healthy human nutrition. A number of studies have reported on the substantial effect which different ratios of saturated and unsaturated fatty acids may exert on the health of those consuming them. While saturated fatty acids are considered a risk factor for cardiovascular diseases (Burr *et al.*, 1989; Hrboticky & Weber, 1993), polyunsaturated fatty acids are regarded as assisting in the prevention of disease (Simopoulos, 1991; Weber *et al.*, 1993; Willett, 1994). Since it was revealed that the fat contained by foodstuffs of animal origin is very rich in saturated fatty acids the popularity of pig and cattle meat products for human consumption has recently suffered a decline, while that of poultry, fish and various sea foods, which contain high levels of unsaturated fatty acids, has increased.

A quantity of information has been published recently in connection with the fatty acid composition and cholesterol content of the back fat and other fat of the Mangalica pig. It has been claimed that the fat of the Mangalica pig is softer and more easily digestible than that of modern pigs. Its softer, granular consistency is attributable to its different, and also healthier, fatty acid composition. Another view expressed is that the cholesterol content of the fat of the Mangalica pig is substantially lower than that of the fat of the new, intensive genotypes. At present the validity of this view can be neither corroborated nor refuted, since, as far as the authors are aware, there are no precise relevant experimental data available. The investigations outlined in this paper were performed for the purpose of providing scientific substantiation or disproval of the above assertions; this study involved the determination of the fatty acid composition and cholesterol content of the fat of Mangalica, Mangalica x Duroc F1 and Hungarian Large White x Hungarian Landrace F1 (MNF x ML) pigs. The MNF x ML genotype is one of the most extensively used crosses in Hungary, and was therefore quite suitable to act as the control.

Materials and method

These investigations were performed with the collaboration of the Hungapig Co. Ltd. and the Animal Breeding and Nutrition Research Institute in Herceghalom, at the new performance testing station established in 1997. The experimental livestock were all housed in the same indoor area, with 6 pigs to a cage and 2.5 m^2 ground area per animal. Throughout the study both the Mangalica pigs and those of the other genotype constructions were fed *ad libitum* diets of identical composition, provided from self-feeders.

At a liveweight of between 120 and 130 kilograms the pigs were slaughtered and their meat classified at the slaughterhouse of the Animal Breeding and Nutrition Research Institute in Herceghalom. After narcosis and slaughter, hanging to drain off the blood, boiling at 60-64 °C and manual singeing away of the hair the carcasses were divided into parts. During the routine splitting and cutting into pieces of the carcasses 100 grams back fat samples were taken from the region of the withers. These samples were stored in a freezer prior to laboratory analysis.

Examination of fatty acid composition and cholesterol content

Both fatty acid composition and cholesterol content was determined by gas chromatography. Conditions applied for gas chromatography:

Equipment: Chrompack CP 9000 gas chromatograph
Column: 50 m x 0.25 mm quartz capillary, humidifying phase CP Sil-88 (FAME)
Detector: FID
Injector: splitter
Gases: carrier gas helium, 150 kPa, rate of flow 30 cm^3/min.; at the detector: air 250 cm^3/min., hydrogen 30 cm^3/min.
Temperatures: injector 220 °C, detector 220 °C, column initially 100 °C, then increasing by 6 °C/min. to 210 °C, and subsequently isothermal until the process was completed
Volume injected: 0.5-2 µl

Statistical evaluation of results

The Student t-test was applied for the statistical evaluation of the experimental data. The analysis of the basic statistics and the correlation analyses were performed by means of the SPSS for Windows (1996) software package, version 7.5.

Results and discussion

Table 1 contains the fatty acid composition of the fat of the pigs of different genotypes in terms of relative mass percentages of the fatty acid methyl esters, while table 2 shows the cholesterol content of the fat of the pigs of the various breeds.

No significant difference (at P = 0.05 level) between the individual genotypes was detected either for unsaturated essential fatty acids or for unsaturated non-essential fatty acids, with the exception of eicosanoic acid. With respect to saturated fatty acids, with the exception of capric, lauric and palmitic acid, difference between the genotypes proved significant at P = 0.05 level. Of these saturated fatty acids, in the case of stearic, margaric, pentadecanoic and nonadecanoic acid the MNF x ML genotype contained the higher

186

proportion, only myristic acid being determined in higher quantities in the Mangalica pig. This signifies that the ratio of saturated fatty acids in comparison with the unsaturated fatty acids was the highest in the MNF x ML pigs (41.12:58.88), although the difference was not significant (this ratio proving to be 39.87:60.13 for the Mangalica). The value for the Mangalica x Duroc genotype was found to be closer to that obtained for the MNF x ML group. It can be ascertained that for every fatty acid under examination the control group differed non-significantly from the Mangalica pigs.

All of the three genotypes included in this study were found to deviate greatly from the hypothetically ideal ratio with respect to fatty acid composition (HIF), ratios for saturated fatty acids being calculated at only approximately 40 % instead of 53-62 %, while those for unsaturated fatty acids proved to be around 60 % rather than 38-47 %. The values for oleic acid (43-44 %) were substantially higher than those reported in the literature, while those for linoleic acid (10-11 %) and those for linolenic acid (0.5-0.7 %) were found to correspond to the literature data.

Table 1.   Fatty acid composition of the fat of the pigs of various genotypes (relative percentage of fatty acid methyl esters)

Fatty acid	Genotype		
	Mangalica, n=5 Mean ± SD	MNF x ML, n=5 Mean ± SD(6)	Mangalica x Duroc, n=5 Mean ± SD(6)
Capric acid	0.071 ± 0.0087	0.08 ± 0.011	0.082 ± 0.0103
Lauric acid	0.09 ± 0.0081	0.084 ± 0.010	0.086 ± 0.0068
Myristic acid	1.64 ± 0.12	1.458 ± 0.116	1.53 ± 0.083
Pentadecanoic acid	0.04 ± 0.0081	0.058 ± 0.012	0.038 ± 0.0062
Palmitic acid	25.97 ± 0.81	25.04 ± 1.01	26.15 ± 0.978
Palmitoleic acid	2.65 ± 0.47	2.27 ± 0.32	2.49 ± 0.424
Margaric acid	0.28 ± 0.034	0.45 ± 0.098	0.262 ± 0.034
Stearic acid	11.56 ± 1.01	13.63 ± 0.698	12.71 ± 1.633
Oleic acid	44.81 ± 1.71	44.34 ± 1.282	43.57 ± 2.155
Nonadecanoic acid	0.059 ± 0.012	0.074 ± 0.019	0.054 ± 0.0049
Linoleic acid	11.47 ± 1.92	10.63 ± 1.609	11.15 ± 0.724
Arachidic acid	0.17 ± 0.017	0.23 ± 0.022	0.2 ± 0.034
Eicosenoic acid	1.02 ± 0.208	0.75 ± 0.095	0.84 ± 0.139
Linolenic acid	0.57 ± 0.042	0.62 ± 0.081	0.63 ± 0.046
Eicosatrienoic acid	0.074 ± 0.0106	0.084 ± 0.022	0.068 ± 0.0091
Arachidonic acid	0.156 ± 0.027	0.196 ± 0.045	0.15 ± 0.021

Table 2   Cholesterol content of the fat of the pigs of various genotypes

Genotype	Cholesterol content (mg/100 g) mean ± SD
Mangalica, n=5	88.4 0 ± 10.08
Hungarian Large White x Hungarian Landrace, n=5	83.60 ± 11.77
Mangalica x Duroc, n=5	92.00 ± 8.72

On the basis of these investigations it may be established that no substantial difference was ascertained with respect to either the monounsaturated, or the polyunsaturated, or the saturated fatty acids (stearic acid being the exception among the fatty acids present in concentrations above 10%) on examination of the fatty acid composition of the fat of these three pig genotypes. In the case of palmitic acid, oleic acid and linoleic acid, which together amount to more than 80 % of fatty acid content, the mean values obtained practically concur. Thus, from these investigations it is possible to draw the conclusion that the fatty acid composition of the fat of the Mangalica pig is, practically speaking, totally identical in value to that of the fat of the Hungarian Large White x Hungarian Landrace and the Mangalica x Duroc genotype constructions. There are therefore no grounds for any assumption that the fat of the Mangalica breed has a more favourable fatty acid composition which would render it more easily digestible and healthier for humans than that of the intensive breeds.

A similar conclusion can be drawn with regard to the cholesterol content of the fat of these genotypes. On the basis of the average for nine animals the cholesterol content of the fat of the Mangalica was measured at 88.44 mg/100 g, that of the Hungarian Large White x Hungarian Landrace at 83.60 mg/100 g, and that of the Mangalica x Duroc F1 genotype at 92.00 mg/100 g. No significant difference at $P < 0.05$ level was detected between the three genotypes with respect to fat cholesterol content; variance within the genotypes proved greater than that between genotypes. Thus, there is no truth in the reports indicating that the fat of the Mangalica pig contains less cholesterol than that of the more generally produced types of fattening pig.

However, on the basis of the findings of these investigations the authors wish to draw attention to the observation that the fat of all three genotypes examined proved to contain 43-45 % oleic acid and 10-12 % linoleic acid, and is thus extremely rich in unsaturated fatty acids and the essential linoleic acid when pigs are kept on a fattening diet based on one of the feed mixes currently in widespread use. The linolenic acid content (0.57-0.63 %) and arachidonic acid content (0.15-0.20 %) of the fat of the pigs examined proved low, while in comparison with the other fats studied stearic acid content was observed to be extremely low (11.56-13.63 %).

The measurements made in this study indicate that the cholesterol content of pig fat varies between 71 and 109 mg/100g. This cholesterol content is substantially lower than that of kidney, liver, egg yolk, bone marrow or cod liver oil.

References

Burr, M.L., A.M. Fehily, J.F. Gilbert, S. Rogers, R.M. Hollidax, P.M. Sweetnam, P.C. Elwood & N.M. Deadman, 1989. Effects of changes in fat, fish, and fibre intakes on death and myocardial reinfarction: Diet and reinfarction trial (DART). Lancet. ii: 757-761

Hrboticky, N. & P. Weber, 1993. Dietary habits and cardiovascular risk. The role of fatty acids, cholesterol and antioxidant vitamins in the prevention and treatment of cardiovascular diseases. In Atherosclerosis, Inflammation and Thrombosis. Neri Serneri G.G, Gensini G.F. R, Abbate R., D. Prisco (editors.), Scientific Press, Florence, 131-152

Simopoulus, A.P. 1991. Omega-3 fatty acids in health and disease and in growth and development. Am. J. Cli. Nutr., 54: 438-463

Weber, P.C., A. Sellmayer & N. Hrboticky, 1993. Fatty acids and their diverse functions: A challenge to future food production. Proceeding, Minisymp., 44th Ann. Meeting EAAP, Copenhagen, 19-27

Willett, W.C. 1994. Diet and health - What should we eat? Science. 265: 532-537

The effect of paternal breed on meat quality of progeny of Hampshire, Duroc and Polish Large White boars

J. Nowachowicz, G. Michalska, B. Rak & W. Kapelanski

University of Technology and Agriculture, Mazowiecka 28, 85-084 Bydgoszcz, Poland

Summary

Three groups, each of 30 gilts being the progeny of Hampshire (H), Duroc (D) and Polish Large White (PLW) boars and PLW sows were analysed. In the samples of *longissimus lumborum* the pH_1, meat colour lightness and soluble meat protein content were determined. The results indicated that meat quality of H x PLW pigs was lowered in respect to water holding ability and in consequence to lowering the technological yield.

Keywords: Pigs, Crossbreds, Hampshire, Meat quality

Introduction

In crossbreeding systems directed on producing the more meaty slaughter pigs the Hampshire and Duroc boars, as the paternal breed, are used. It is well documented, however, that improved carcass lean content may be the reason for meat quality deterioration (Rozycki & Tyra, 1998; Steinhauf *et al.*, 1976). An additional disadvantage with Hampshire pigs is "acid meat" which bring about a decrease of technological yield (Naveau, 1986; Lundstrom *et al.*, 1996; 1998).

The aim of the study was to compare some essential meat quality traits in crossbred pigs from Hampshire, Duroc and Polish Large White boars and Polish Large White sows.

Material and methods

The study was carried out on three groups of gilts being the progeny of Hampshire, (H), Duroc (D) and Polish Large White (PLW) boars and Polish Large White sows. Each group consisted of 30 gilts reared in the Hybridisation Centre in Pawlowice belonging to National Institute of Animal Production. Pigs were slaughtered at 185 days of age.

Meat quality traits were determined on the *longissimus lumborum* muscle according to methods used in Progeny Testing Stations (Kielanowski *et al.*, 1977). The pH_1 value was measured 45 minutes after slaughter. Twenty four hours later the meat colour lightness was measured with the use of leukometer, and meat soluble protein content as described by Kotik (1974).

The data were statistically analysed using the usual procedure of one-way variance analysis with respect to crossbred groups (Ruszczyc, 1978).

Results and discussion

Mean values of meat quality traits are listed in Table 1. Acidity of meat measured at 45 minutes after slaughter (pH_1) was the same in the compared pig groups and amounted to 6.32 in H x PLW, 6.33 in D x PLW and 6.31 in purebred PLW. In an earlier study Michalski *et al.*,

(1988; 1990) has shown mean pH_1 values 6.55 for crossbred H x PLW pigs and 6.21 for PLW. Other data showed a higher pH_1 value 6.61 for PLW pigs (Fandrejewska, 1997).

Table 1. Meat characteristics.

Trait	Groups		
	H x PLW	D x PLW	PLW x PLW
pH_1	6.32 ± 0.16	6.33 ± 0.14	6.31 ± 0.39
Colour lightness, %	24.66 ± 2.39	24.54 ± 1.31	23.61 ± 2.30
Meat soluble protein content, %	7.93[Ab] ± 0.33	8.25[bc] ± 0.40	8.73[Ac] ± 0.69

Means with the same letters b, c: significant at $P < 0.05$; A: significant at $P < 0.01$

Muscle colour lightness did not show any significant differences between the compared pig groups. Similar data has been obtained by many authors (Michalski *et al.*, 1988; 1990; Fandrejewska, 1997). It should be stressed that both pH_1 and colour values obtained in the present study were in the range ascribed to normal meat quality.

   With regard to the meat soluble protein content there have been significant differences among compared pig groups. The lowest level was in H x PLW crosses (7.93%) in comparison with D x PLW (8.73%) at $P < 0.05$ and with purebred PLW (8.73%) at $P < 0.01$. Low level of water soluble protein in meat is characteristic of meat with PSE symptoms (Grajewska *et al.*, 1984) or with higher denaturation of native meat protein (Kotik, 1974) and therefore with a lower water holding capacity of meat. It is evident from many reports that in pure Hampshire or their crosses the technological meat quality is lowered (Lundstrom *et al.*, 1996; 1998; Monin & Sellier, 1985; Naveau, 1986; Sellier & Monin, 1994). Our result, in some degree, has proved the poorer usefulness of Hampshire boars for slaughter pig production.

Conclusion

1. The most favourable level of soluble meat protein content was in PLW pigs and the lowest in Hampshire crossbred pigs.
2. In both pH_1 and colour lightness values no significant differences were shown among the H x PLW, D x PLW and purebred PLW pigs.

References

Fandrejewska, M., 1997. Evaluation of meat quality in large white x hampshire crossbreed pigs. Polish-Slovak-Czech Scientific Meeting "Current problems on pig production". Olsztyn, 98

Grajewska, S., J. Kortz & J. Rozyczka, 1984. Estimation of the incidence of PSE and DFD in pork. Proc. Scient. Meeting. Biophysical PSE-muscle analysis. Vienna, 72-89

Kielanowski, J., H. Duniec, T. Kostyra, M. Kotarbinska, F. Maly, Z. Osinska, M. Rozycki & W. Szulc, 1977. [Results of evaluation in pig testing stations of the Institute of Zootechnics in 1976]. PWRiL, Warszawa, 5-28

Kotik, T., 1974. [Protein content in muscle extracts as an index of meat quality]. Rocz. Inst. Przem. Miesnego 9: 47-52

Lundstrom, K., A. Andersson & I. Hansson, 1996. Effect of the RN⁻ gene on technological and sensory meat quality in crossbred pigs with Hampshire as terminal sire. Meat Sci. 42: 145-153

Lundstrom, K., A.C. Enfalt, E. Tornberg & H. Agerhem, 1998. Sensory and technological meat quality in carriers and non-carriers of the RN⁻ allele in Hampshire crosses and in purebred Yorkshire pigs. Meat Sci. 48: 115-124

Michalski, Z., D. Ceglarska & M. Kamyczek, 1988. [Comparison of the quality parameters of pork muscle in pure breed and in crosses]. Zesz. Probl. Post. Nuk Rol. 335: 23-27

Michalski, Z., D. Ceglarska & M. Kamyczek, 1990. [Physico-chemical evaluation of meat of the Duroc and Polish Large white pigs]. Zesz. Probl. Post. Nauk Rol. 384: 109-113

Monin, G. & P. Sellier, 1985. Pork of low technological quality with a normal rate of muscle pH fall in the immediate post-mortem period: the case of the Hampshire breed. Meat Sci. 13: 49-63

Naveau, J., 1986. Contribution a l'etude du determinisme genetique de la qualite de viande porcine. Heritabilite du rendement technologique Napole. J. Rech. Porcine en France. 18: 265-276

Rozycki, M. & M. Tyra, 1998. Wyniki oceny uzytkowosci tucznej i rzeznej swin w stacjach kontroli. Stan hodowli i wyniki oceny swin w roku 1997. Krakow 1998, 75-84

Ruszczyc, Z., 1978. [Statistical methods in animal experiments]. PWRiL, Warszawa

Sellier, P. & G. Monin, 1994. Genetics of pig meat quality: A review. J. Muscle Foods, 5: 187-219

Steinhauf, D., J.H. Weniger & H.P. Mader, 1976. Observations on the apparent antagonism between producing capacity and meat quality. In Meat animals growth and productivity. Eds. D. Lister, J.N. Rhodes, V.R. Fowler & M.F. Fowler. Plenum Press, New York, 373-385

Comparison of several pig breeds in fattening and meat quality in some experimental conditions of a Czech region

T. Adamec, B. Naděje, J. Laštovková & M. Koucký

Research Institute of Animal Production, Praha 10 Uhříněves, 104 01, Czech Republic

Summary

Experimental fattening of LW, LWxL, (LWxL)xH, (LWxL)x(LW imported), (LWxL)x(BLxD) and (LWxL)xCMP (Czech Meat Pig) breeds was done in 1992-1998. These groups, apart from the LW were involved in more than 2 repetitions. Fattened young boars were tested in the last groups of breed, too. Conditions of fattening, feedstuff composition, individual boxing, transport, slaughtering and laboratory methods were identical.

The (LWxL)xH breed was the best in growing intensity (789g), feed consumption (3.21kg/kg) and in high lean meat (55%) followed by the (LWxL)x(BLxD) breed (816g, 3.11kg/kg respectively), with low lean meat (49.6%) however, which was similar to the original farm condition. The content of intramuscular fat in m.l.l.t. was lower in hybrids of LW (imported), BL and D breeds (1,3-1,7% unlike 1,7-1,9%), all in sire positions.

The (LWxL)xCMP breed was best in sensory evaluation in texture and humidity and the LW and (LWxL)H breeds in flavour and taste of m.l.l.t.

The young boars of the (LWxL)xCMP breed were absolutely best in daily gains, feed consumption and lean meat, however, were lowest in the content of dry matter, protein and intramuscular fat of m.l.l.t. (1,1%) and also the lowest in the evaluation of all sensory criteria.

Keywords: Pigs, Breed, Meat quality, Sensorial evaluation

Introduction

Comparison of the feeding criteria, namely meatiness, meat quality and sensorial evaluation of several pig breeds and young boars in the conditions of the Czech Republic was the aim of this presentation. This work is the comparison in which, unlike various conditions in practise, identical conditions of housing, nutrition and slaughtering were applied.

Analytical and sensorial evaluation including young boars completed the comparison of the qualitative parameters.

Material and methods

Fattening of pigs in groups according to their breeds or hybrids was done in experimental conditions from 1993 to 1998. Piglets came out of various breedings. Gilts, castrates and young boars were fed *ad libitum* with dry matter mixtures of the same composition in two phases up to 65 kilograms of their liveweights and to the very end of fattening. Individual stabling, transportation, method of slaughtering and analyses were identical. Analytical determinations were done in the RIAP laboratories by the established methods. Determination of a meat share was done by a two-point measurement method of the EUROP system by means of regressive equations valid in the Czech Republic.

The sensorial evaluation of the m.l.l.t. samples of both sexes was done by a ten-member commission. Owing to a gradual accumulation of the gained results, a method of a point evaluation and average determination was chosen. Some hybrids were tested using a lower number of representative samples. Points from one (the worst point) to seven were assigned to the quality of flavour, taste, humidity and texture.

Results and discussion

The results are shown in tables 1 and 2. The best daily gains, feed conversion and higher meat share were found in the LW breed. High daily gains and the lowest feed consumption could be due to a low number of cases, a result of incorrect breeding of the gene pool for the maternal position. In the Czech Republic there is a type of the LW*m (maternal position) and the LW*p (paternal position) breeds. These values were low at the maternal combination of the LWxL crossing. The results, however, cannot be related to five hybrids at which the maternal combination (LWxL) always comes out from a different breeding.

Table 1. Selected evaluation criteria.

criteria:		LW	LWxL	(LWxL)xH*	(LWxL)xLW	(LWxL)x (BLxD)	(LWxL)xCMP	(LWxL)x CMP boars
	n=	7	34	53	26	49	72	18
slaughter weight	x	103	110	104	103	105	111	106
	s	14	10	3	3	4	9	4
start weight	kg	27	24	22	24	29	28	28
	s	3	4	2	3	5	9	6
daily gain kg		908	757	789	781	816	720	891
	s	38	112	58	51	74	60	121
consumption kg/kg		2.78	3.52	3.21	3.57	3.11	3.75	3.24
	s	0.37	0.60	0.28	0.07	0.28	0.37	0.22
lean meat %		51.4	48.4	55.1	55.3	49.6	51.9	57.3
	s	3.9	2.8	3.5	2.9	3.5	4.1	3.1
intramuscular fat g.kg^{-1}		1.65	1.96	1.67	1.25	1.69	1.84	1.12
	s	0.51	0.79	0.55	0.48	0.70	0.77	0.23
dry matter mllt g.kg^{-1}		257	263	250	253	259	259	248
	s	9	10	7	4	7	8	4
protein g.kg^{-1}		226	229	214	226	227	225	221
	s	5	6	9	3	5	8	6
NPV / Intr.fat		13.0	11.8	12.3	18.2	13.8	11.9	17.2
	s	4	5	4	8	6	4	4

As expected the best daily gains and meat share were found in the final hybrids of young boars (LWxL)xCMP /Czech Meat Pig/ at a very good feed conversion. These values were the worst at the (LWxL)xCMP hybrid. In the combination of these criteria the best results were achieved by the

(LWxL)xH final hybrid. This phenomenon can also be seen in operating practice.

The majority of the gained statistical data on the intensity of growth, feed conversion and meat share of the breeds came from the conditions tested. During the tests, nutrition is being ensured by a highly valuable nutritive mixture for the entire feeding time and the tested piglets were selected by their age and weight to reach good results. According to the results taken from the testing stations in 1998 the values of meat share in the LW*m breed are 54.9% (from 51,8% to 57,5%) and the daily gain is 895 g (from 829 g to 946 g) at a feed consumption from 2,35 to 3,04 kg/kg of gain.

Indexes of daily gains and feed consumption are only preliminary owing to the different average weight of pigs at the beginning of the experiment. Values of meatiness and meat quality are, however, comparable.

These values are of the same scale of the other breeds (LW*p, L*m, CMP). These results are not being obviously reached under the current operating conditions.

In the tests the nutritive level of nutrition and the selection of piglets were the same as in operating practice. In the (LWxL)x(BLxD) hybrid the results were not typical for low meat share but they were of the same value as at the original state. This state was sufficient owing to a reasonable amount of money received for the fat. The average meat share of the hybrids was 52,98% and this value corresponds to operating conditions - 52,34% according to Matoušek et al. (1997) and 51,75% according to Adamec and Naděje (1998) in the amount of 2,9% at the (LWxL)xCMP hybrid with its slaughter weight of 115 kilograms.

Table 2. Averages of point sensorial evaluation (the best point is 7 - maximum).

criteria:		LW	LWxL	(LWxL)xH	(LWxL)xLW*	(LWxL)x(BLxD)	(LWxL)xCMP	(LWxL)xCMP boars
n=		7	34	13	6	10	72	18
flavour	x	6.1	5.6	5.9	5.7	5.3	5.4	4.8
	s	0.37	0.68	0.33	0.06	0.19	0.61	0.52
taste		5.9	5.5	5.6	5.6	5.1	5.2	3.9
		0.37	0.70	0.37	0.15	0.41	0.76	0.30
humidity		5.1	4.9	4.8	4.9	4.6	5.2	4.1
		0.40	0.60	0.72	0.21	0.52	0.44	0.67
texture		5.4	5.2	5.0	5.2	4.7	5.5	4.2
		0.57	0.70	0.89	0.28	0.80	0.55	0.60

s= standard deviation

In the tested groups there was an extremely low content of intramuscular fat in the (LWxL)x LW* (*= imported) and also in the young boars of the (LWxL)xCMP hybrid. As a whole, the content of intramuscular fat was very low for the average values and thus the same problem appears in the Czech Republic as in some Western European countries.

Significant differences were not found among the breeds as far as dry matter and proteins are concerned. More significant differences were found in the ratio of the NVP (nutrition protein value): the content of intramuscular fat corresponding to the contents of intramuscular fat.

The results of sensorial measurement should be handled separately. The best results were found for the LWm breed for the maternal combination of crossing. From the final hybrids the (LWxL)xCMP hybrid was the best evaluated for texture and humidity and the (LWxL)xH hybrid for flavour and taste. The young boars were evaluated in the lowest position in all criteria but boar smell of meat was recorded just in two samples from the whole number. As a whole the meat of young boars was evaluated as acceptable. From previous knowledge Adamec (1998) pointed out a narrow relationship between flavour and taste and between humidity and texture, and a low relationship among them. Only a low correlation to the content of intramuscular fat, slaughter weight and time of fattening was evaluated for flavour and taste.

Conclusion

Fattening ability, carcass value and meat quality characteristics found in this study are on usual levels, except for the low content of intramuscular fat in m.l.l.t., even though differences by some specific effect, especially by different conditions during the period of piglets rearing. In the hybrid (LWxL)xCMP, the group of gilts and castrates was compared with group of young boars. All differences were statistically significant.

References

Adamec,T. & B. Naděje, 1998. Performance, carcass value and meat quality of the final hybrid (LWxL)xCzech Meat Pig, the most frequent type in the Czech Republic. Book of abstracts of EAAP-49th, Warsaw: 259

Matoušek, V., N. Kernerová, J. Václavovský & A. Vejčík, 1997. Analysis of meat quality in hybrid population of pigs. Czech J. Anim. Prod. Sci. 42: 511-515

Anonymous, 1998. Pig performance. A report of the Union of pig breeders in Bohemia and Moravia and The Bohemian and Moravian association of breeders, ltd. p.193

Criteria of pork quality

Fat deposition and distribution in three genetic lines of pigs from 10 to 105 kilograms liveweight

K. Kolstad

Department of Animal Science, Agricultural University of Norway, P.Box 5025, N-1432 Ås

Summary

The object of this study was to investigate genetic differences in fat amounts and distribution from weaning to slaughter at 105 kilograms. Fat deposition was investigated in 141 pigs of the three genetic groups Norwegian Landrace (L), Duroc (D) and a cross between Norwegian Landrace and a Norwegian Landrace line selected for increased backfat and slow growth (LP*L). Amounts and proportions of the depots subcutaneous, inter/intra muscular and internal fat were quantified at five separate times in each animal during growth by computer tomography (CT). There were breed differences in fat deposition and distribution

Magnitude of breed differences in fat depots varied between the fat depots; exterior depots showed larger breed differences than the interior fat depots. Breed differences were present at lower liveweights. Higher amounts and proportions of internal fat in the Landrace were found at 25 kilograms liveweight. Differences in subcutaneous fat between LLP and the modern breeds was also found at this stage. Duroc had a significant effect on amounts and proportions of inter/intramuscular fat from 85 kilograms liveweight. This suggests that breeding for differences in fat distribution can be based on registrations on young pigs.

Keywords: Fat deposition, Fat distribution, Breed differences

Introduction

Earlier studies based on CT observations of different breeds of pigs by Luiting *et al.* (1995), Kolstad & Vangen (1996) and Kolstad *et al.* (1996) proved breed differences in maintenance requirements, fat distribution and fat mobilisation at about 60 kilograms liveweight, i.e. when the pigs grow most rapidly.

The aim of the present study was to examine development of different fat depots in breeds of pigs, and to prove whether there are breed differences in fat distribution as well as in amounts of fat depots at different stages of growth by repeated measures of body composition by CT within animals from weaning to slaughter.

Material and methods

Animals: In total 141 pigs were included in the experiment representing the three breeds/lines Landrace (63), Duroc (62) and L*LP(16) which is a cross between Norwegian Landrace (L) and a Norwegian Landrace (LP) selection line selected for increased backfat and slow growth over 8 generations (Vangen, 1979).

Computer tomography: Computer Tomography (CT) was used for measuring body compositional changes during the experiment. The pigs were scanned at weaning (~10 kg), at about 25, 50, 85 and 105 kilograms liveweight, in total five times for each animal. They were fasted for more than 16 hours before scanning. Fat depots were quantified. The sum of all

199

tissues observed by CT in each animal is referred to as CT-weight. Technical details about computer tomography are described in detail in Kolstad & Vangen (1996).

Statistical analysis: Estimates and tests of fixed (breed, sex, batch, liveweight, total fat) and random effects (animal, error) were based on a REML analysis using an animal model using the ASReml software developed by Gilmour *et al.* (1997). Interactions were included in the model whenever they were found significant.

Allometric growth coefficients were estimated within each animal relative to CT-weight and total fat based on the same model combined with log transformation of the dependent variable and CT- or live weight.

Results

Subcutaneous fat is the earliest maturing fat depot while internal and inter/intramuscular fat are later maturing depots in all three genetic groups (Table 1).

Table 1. Allometric growth coefficients for fat growth from 10 to 105 kg liveweight. Y=fat depots, X=CT-weight or total fat.

	W=CT-weight			W=total fat		
Fat depot:	Landrace	Duroc	LLP	Landrace	Duroc	LLP*
Subcutaneous	1.00	1.02	1.19	0.85	0.80	0.84
Inter/intramuscular	1.44	1.58	1.71	1.25	1.21	1.23
Internal	1.27[a]	1.38[ab]	1.46[b]	1.09	1.12	1.06

[a,b] Indicate significant differences between breeds within fat depot
* Crossing between Landrace and a selection line selected for increased backfat and slow growth.

Both inter/intra muscular fat and internal fat increase in amounts relative to CT-weight (b>1.0), while subcutaneous fat increased at about the same rate as CT-weight. The differences between the breeds are relatively larger at higher than at lower liveweights (Figure 1). Generally, Landrace have lower amounts of subcutaneous and inter/intramuscular fat than the Duroc, but more internal fat when adjusting for CT-weight.

The differences between LLP and the two modern breeds in amounts of fat adjusted for CT-weight increases with increasing liveweight, and are largest for subcutaneous fat, less for inter/intramuscular fat and least for internal fat (Figure 1).

Subcutaneous fat decreases in proportion to total fat on behalf of inter/intramuscular fat, as the allometric growth coefficient for subcutaneous fat is less than 1.0, above 1.0 for inter/intramuscular fat and close to 1.0 for internal fat when estimated relative to total fat. The changes in relative proportion of fat depots during growth were equal for all three breeds (Table 1, Figure 2).

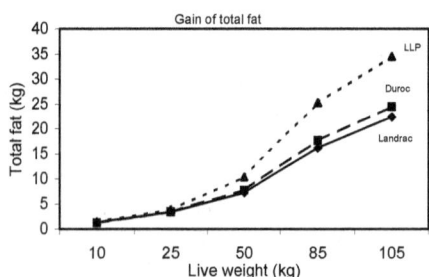

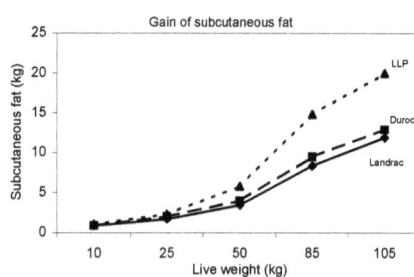

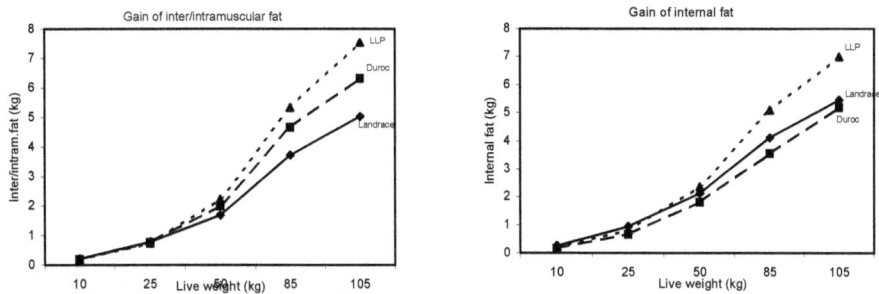

Figure 1. Deposition of fat totally, and in the subcutaneously, inter/intramuscular and internal fat depot from 10 to 105 kg live weight in Landrace, Duroc and LLP* pigs.

When adjusting amounts of each fat depot to equal amounts of total fat, significant breed differences appear (Table 2, Figure 2). The subcutaneous fat depot is the dominating fat depot at all stages. Its proportion is however highest at 10 kilograms liveweight accounting for 65-75 percent of total fat amount. The biggest change in distribution of fat between the depots is from 10 to 25 kilograms liveweight, where all breed decrease their proportion of subcutaneous fat, and increase the proportions of the two other depots.

Table 2. Breed differences in amounts or proportion of each fat depot from 10 to 105 kg liveweight when adjusting to ct weight or total fat. (CT-weight=sum of body components measured by CT, Total fat=sum of fat components).

		Adjusted for CT-weight			Adjusted for Total fat		
Live weigh	Fat depots	D - L	LLP - L	LLP - D	D - L	LLP - L	LLP - D
10 kg	Subcutaneous	0.012	0.148	0.136	0.071**	0.077**	0.006
	Inter/intramuscular	-0.017	-0.014	0.003	0.025	0.007	-0.018
	Internal	-0.087*	-0.046	0041	-0.097**	-0.087*	0.010
25 kg	Subcutaneous	0.294	0.591*	0.297	0.243*	0.440**	0.197
	Inter/intramuscular	-0.010	-0.045	-0.036	-0.144*	-0.197^{t}	-0.053^{t}
	Internal	-0.278**	-0.131^{t}	0.147^{t}	-0.314**	-0.329**	-0.015
50 kg	Subcutaneous	0.548^{t}	2.358**	1.81**	0.382^{t}	0.729**	0.347
	Inter/intramuscular	0.307	0.532^{t}	0.225	0.198	-0.139	-0.337
	Internal	-0.298**	0.244^{t}	0.542**	-0.487***	-0.388**	0.099
85 kg	Subcutaneous	1.145**	6.500***	5.355***	0.372^{t}	1.191***	0.819**
	Inter/intramuscular	0.941**	1.611***	0.67*	0.685***	-0.244	-0.929***
	Internal	-0.563**	0.977**	1.54***	-0.818***	-0.605***	0.213
105 kg	Subcutaneous	1.044^{t}	8.084***	7.040***	0.113	1.612***	1.499***
	Inter/intramuscular	1.289**	2.507***	1.218**	0.922***	-0.356	-1.278***
	Internal	-0.264	1.526***	1.790***	-1.002***	-0.724***	0.278

^{t,*,**} and *** indicate differences at significant levels p=0.10, p=0.05, p=0.01 and p=0.001

Landrace have a relatively high proportion of internal fat when adjusting for total fat during the whole experimental period. Duroc have a high proportion of inter/intramuscular fat and LLP a high proportion of subcutaneous fat (Figure 2, Table 2).

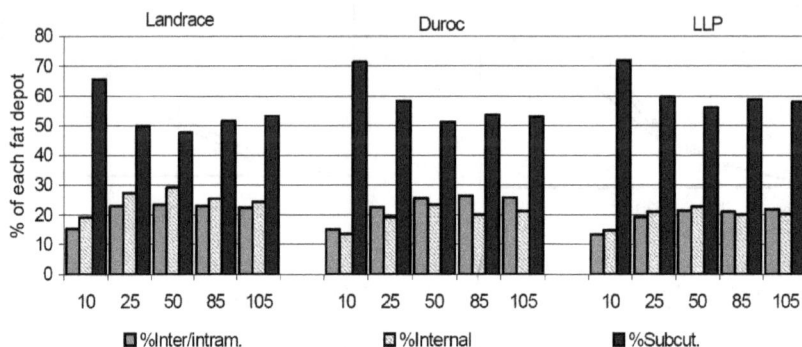

Figure 2. Fat distribution on the three depots subcutaneous, inter/intramuscular and internal fat in Landrace, Duroc and LLP pigs at 10, 25, 50, 85 and 105 kg live weight.

Discussion/conclusion

Results from the present study indicate:
- There are breed differences in fat deposition and distribution
- Magnitude of breed differences in fat depots varied:
  Subcutaneous>inter/intramuscular>Internal
- Differences are present at lower liveweights (Internal fat in the Landrace, Subcutaneous fat in the LLP, Inter/intramuscular fat in the Duroc)
- High amounts and proportions of internal fat in the efficient, fast growing Landrace breed during growth
- Marbling, the Duroc effect, appears close to sexual maturity (85 kg)
- Modern breeds do not differ in fat amounts from 'old' types at weaning, but are large at slaughter
- Subcutaneous fat depot is the largest fat depot, but is relatively largest at weaning
- CT a powerful tool for studies of growth and development as within animal variation is accounted for

References

Gilmour, A. R., R. Thompson, B. R. Cullis & S. J. Welham, 1997. ASREML

Kolstad, K., N. B. Jopsen & O. Vangen, 1996. Breed and sex differences in fat distribution and mobilisation in growing pigs fed at maintenance. Livest. Prod. Sci. 47: 33-41

Kolstad, K. & O. Vangen, 1996. Genetic differences in maintenance efficiency when accounting for changes in body composition. Livest. Prod. Sci. 47: 23-32

Luiting, P., K. Kolstad, H. Enting & O. Vangen, 1995. Breed comparison of body composition in pigs fed at maintenance. Livest. Prod. Sci. 43: 225-234

Standal, N., E. Vold, O. Trygstad & I. Foss, 1973. Lipid mobilisation in pigs selected for leanness or fatness. Anima. Prod. 16: 37-42

Vangen, O.,1979.Studies on a two trait selection experiment in pigs. II. Genetic changes and realised genetic parameters in the traits under selection.Act.Agric.Scand.29:305-319

Fat score, an index value for fat quality in pigs – its ability to predict properties of backfat differing in fatty acid composition

K. R. Gläser, M.R.L. Scheeder & C. Wenk

Swiss Federal Institute of Technology (ETH) Zurich, Institute of Animal Science, Nutrition Biology, CH – 8092 Zurich, Switzerland

Summary

Fat score, a semi-automated determination of double bonds in adipose tissue, is an established method in Swiss slaughterplants to assess backfat quality in pigs. Since body fat composition is directly influenced by dietary fat, two feeding trials (i and ii) were conducted. In the first trial i) the effects of monoenoic or polyenoic fatty acids at similar number of double bonds per kilogram feedstuff (70, 49.5, 31.7 g/kg pork fat, olive oil, soybean oil, respectively) on fat score, consistency and oxidative stability were studied. In a second trial ii) the effects of olein and stearin fraction of pork fat and hydrogenated fat (50 g/kg) were investigated.
Olive and soybean oil impaired fat score and firmness of adipose tissue to a greater extent than pork fat supplemented feed. In contrast, highest induction times and therefore best oxidative stability was measured in fat from pigs fed olive oil. The hydrogenated fat resulted in lower fat scores and higher firmness. Compared to the stearin fraction, the olein fraction led to slightly higher fat scores, softer consistency and reduced oxidative stability, but without reaching statistical significance. Significant correlations were found between fat score and consistency (i: $r2=0.672$) and oxidative stability (i: $r2=0.280$, ii: $r2=0.166$). It may be concluded that fat score gives an useful at-line estimate of firmness of pig adipose tissue.

Keywords: Fat quality, Pigs, Fatty acids, Adipose tissue, Fat score

Introduction

The properties of meat and fat tissue in pigs are important for production and quality of meat products and are largely influenced by feeding and genetic disposition of the animals. Due to the successful breeding for lean carcasses the relative amount of polyunsaturated fatty acids (PUFA), which are efficiently deposited into adipose tissues, is high in modern pigs. High amounts of PUFA lead to increased susceptibility to oxidation and, like monoenoic fatty acids (MUFA), impair consistency.

To avoid an unacceptable deterioration of fat quality the fat score – a semi-automated determination of double bonds in adipose tissue – was established in Swiss slaughterplants as an at-line method to assess backfat quality in pigs.

The purpose of the present study was to investigate the influence of different dietary fat sources on to the properties of pig adipose tissue and to focus on the relationship between fat score, oxidative stability and consistency.

Material and Methods

In each of two feeding trials 12 x 4 siblings of Large White (LW) and Swiss Landrace (SL) breed were allocated to 4 feeding treatments, fully balanced according to progeny group, sex and initial weight and fattened from 24 ± 4.9 kilograms to 104 ± 3.6 kilograms liveweight. The pigs were housed in groups of four animals and had *ad libitum* access to feed and water. The basic diets (CI, CII) of both trials were composed of barley, wheat and soybean meal as well as amino acid and mineral additives. The amount of fat supplemented to the feeds of the first feeding trial was calculated to achieve equivalent amounts of double bonds per kilogram of feed: 7% pig fat (PF), 4.95% olive oil (OO), 3.17% soybean oil (SO). In the second trial, equivalent amounts of fat (5%) were added, but with different proportions of saturated and unsaturated fatty acids due to fractionation of pig fat – olein (OLE), stearin (STE) – and hydration (Satura, SAT).

Backfat was taken after carcass dissection (30 h *post mortem*) for the determination of oxidative stability, fat score and consistency. Determination of fat score was carried out with a titro-processor (Metrohm, Schweiz) according to the established method (Scheeder *et al.*, 1999). Fatty acid profile was determined with a gas chromatograph after extraction of lipids from tissue with dichlormethan / methanol (2:1) and derivatisation of fatty acids to fatty acid methylesther.

Oxidative stability was measured as induction time with a Rancimat (Metrohm, Herisau, Schweiz) at 110°C and 20 l / h air flow. Fat consistency was determined with a texture analyser (TA-XT2, Stabel Micro Systems, Haslemere, Surrey, U.K.) in extracted fat (dichlormethan / methanol, 2:1) of the outer layer of backfat by measuring the penetration force of a punch (ø 3.5 mm) at 3 °C and a penetration depth of 10 mm.

Statistical analysis was carried out by GLM procedure and Scheffé's multiple comparison (SAS, release 6.12) regarding the feeding treatment, breed, litter within breed, sex and their interactions as fixed factors and the final live weight as a covariate.

Results and Discussion

The supplementation of different fats had no effect on growth performance and carcass composition except for SAT, which led to decreased weight gain and slightly leaner carcasses, probably caused by a decreased digestibility of this type of fat (Bee *et al.*, 1995). The composition of adipose tissue as well as fat score, oxidative stability and consistency were significantly influenced by the fatty acid composition of the dietary fat.

The high amount of saturated fatty acids in adipose tissues of control animals was most likely the result of increased *de novo* fatty acid synthesis. Although the fat enriched diets of the first trial contained similar amounts of double bonds per kilogram of feed, both plant oils led to a significant increase of MUFA and PUFA, respectively. While monoenoic fatty acids from the pig fat (PF, OLE, STE) and olive oil supplemented feed were deposited at the expense of the saturated fatty acids (SFA), high amounts of PUFA in the SO diet led to a low proportion of MUFA. This may result from a simple relative displacement of monoenoic by polyenoic fatty acids. But, as the proportion of SFA was not as strongly affected as MUFA it can be assumed that *Δ9-desaturase* was inhibited by the high amounts of linoleic acid from the soybean oil (Kouba und Mourot, 1997). Between the diets with the fractionated pig fat no significant differences in fatty acid composition were found, although STE feed was about 10 % higher in SFA. It can therefore be assumed that the MUFA content in adipose tissue is dependent on regulatory mechanisms to a large extent and can be markedly increased by feeding only when a very high amount of MUFA is fed or, most probably, when a specific triglyceride composition supports a preferred deposition into adipose tissue. Olive oil is not only very rich in oleic acid but oleic acid occupies also more than 80 % of the central (*sn*-2) position in its triglycerides. Fatty acids at this position may be well protected from the specific hydrolytic action of pancreatic and lipoprotein lipases and might therefore be readily deposited and kept in the adipose tissues.

Table 1. Fatty acid composition in feedstuff and adipose tissue as well as growth performance, carcass composition and backfat quality in pigs fed diets containing different fat supplements.

	Experiment I				Experiment II			
	CI control I	PF pig fat	OO olive oil	SO soybean oil	CII control II	OLE olein	STE stearin	SAT satura
Feedstuff								
SFA[1) [g/kg]	3.3	30.4	10.2	7.9	3.4	19.4	30.7	32.8
MUFA[2) [g/kg]	2.8	31.4	38.2	10.8	3.0	28.7	24.0	10.7
PUFA[3) [g/kg]	8.8	16.7	15.1	25.7	8.9	16.4	13.8	10.0
DB[4) [mol/kg]	0.07	0.23	0.24	0.22	0.07	0.22	0.18	0.11
Performance								
Weight gain [g/d]	889	867	862	824	910^a	917^a	880^{ab}	802^b
CW [kg][5)	82.3	82.5	83.0	82.3	81.2	81.2	82.3	81.4
Premium cuts [%][6)	55.4	55.3	55.3	54.1	55.4	55.1	56.0	57.2
Fat tissue [%][7)	14.0	13.5	13.3	14.6	13.1	13.7	13.0	12.7
Backfat								
SFA [%]	38.7^a	35.1^b	29.7^c	34.3^b	38.6^a	35.0^b	36.2^b	39.5^a
MUFA [%]	46.5^c	49.8^b	56.2^a	39.6^d	48.0^b	50.7^a	50.3^a	48.5^b
PUFA [%]	14.9^b	15.2^b	14.2^b	25.7^a	13.1^a	14.0^a	13.1^a	11.7^b
DB [mol/kg]	2.67^c	2.81^{bc}	2.94^b	3.24^a	2.59^b	2.75^a	2.67^a	2.52^b
C 16:0	23.4^a	21.3^b	19.3^c	21.0^b	23.5^a	21.9^b	22.5^b	22.3^b
C 16:1	2.6^a	2.7^a	2.2^b	1.9^b	2.6^{ab}	2.8^a	2.8^a	2.4^b
C 18:0	13.3^a	11.9^b	8.8^c	12.1^{ab}	13.3^b	11.3^c	12.0^c	15.2^a
C 18:1	42.5^c	45.6^b	52.7^a	36.6^d	43.9^b	46.2^a	45.9^a	44.7^b
C 18:2	12.2^b	12.3^b	11.9^b	21.3^a	10.5^a	11.3^a	10.6^a	9.6^b
C 18:3	1.1^b	1.2^b	1.1^b	2.1^a	1.1^b	1.2^a	1.1^{ab}	0.9^c
Fat score	60.8^c	64.2^b	66.8^b	70.1^a	59.4^{bc}	62.1^a	61.3^{ab}	57.6^c
Consistency [g]	150^a	113^b	52^c	50^c	181^b	158^b	174^b	279^a
Induction time [h] [8)	4.4^b	4.2^b	4.6^b	2.5^a	4.3^a	4.2^a	4.6^a	4.6^a

[1) saturated, [2) monoenoic, [3) polyenoic fatty acids, [4) Double bonds, [5) Carcass weight (hot), [6) Trimmed shoulder, loin and ham as percentage of carcass weight, [7) Fat and hide trim of premium cuts, [8) measured with Rancimat

LS-Means in one row and within experiment lacking a common superscript are significantly different (P≤0.05)

According to the fatty acid composition, highest fat scores were measured for the soybean oil and olive oil treatment and lowest in pigs fed the hydrogenated fat. The more unsaturated olein led to only slightly higher fat scores and softer consistency than stearin. Although adipose tissue of the control group was nearly as high in SFA content as SAT, consistency was significantly lower. This may be explained by the higher proportion of C 16:0 in the SFA of control in combination with the higher proportion of PUFA. For the olive and soybean oil treatment a remarkable decrease of backfat consistency was found, which clearly underlined that consistency is impaired by high amounts of PUFA as well as MUFA (Suomi *et al.*, 1993). However, the deviation from a linear relationship between fat score and consistency can be mainly led back to SO treatment and it may be assumed that the effect on consistency may be overestimated by the fat score when it is high mainly due to a high PUFA content. On the other hand, consistency of both plant oil treatments was on a level probably too low to detect any differences and it cannot be ruled out that at a lower temperature, consistency of the two treatments would differ to some extent. Overall, a close correlation ($r^2 = 0.78$) was found between fat score and consistency and it can therefore be concluded that fat score gives a useful at-line estimate of firmness of pig backfat.

Figure 1. Correlation between fat score and consistency as well as fat score and oxidative stability.

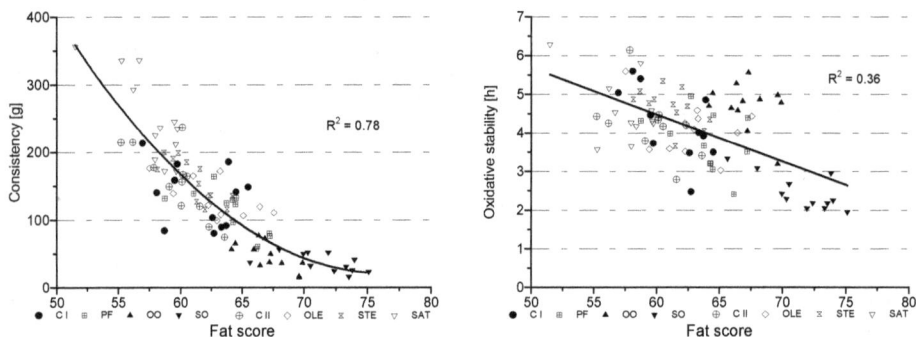

On the other hand oxidative stability showed a lower correlation ($r^2 = 0.36$) with the fat score. While high amounts of PUFA (soybean oil) decreased oxidative stability, olive oil led to slightly higher induction times. This could be explained by the higher stability of MUFA but can possibly also be attributed to the antioxidative activity of phenols which are constituents of olive oil (Baldoni *et al.*, 1996). Obviously, oxidative stability is largely influenced by content and activity of anti- and prooxidative factors beside fatty acid composition (Flachowsky *et al.*, 1997).

References

Bee, G., A. Zimmermann, R. Messikommer, & C. Wenk, 1995. Einfluss des Fettsäurenmusters auf die Verdaulichkeit von Fettzulagen im Magendarmtrakt beim Schwein. J. Anim. Physiol. a. Anim. Nutr. 73: 258-268

Baldoni, M. et al., 1996. Antioxidant activity of tocopherols and phenolic compounds of virgin olive oil. JAOCS 73: 1589-1593

Flachowsky, G., F. Schöne, G. Schaarmann, F. Lübbe & H. Böhme, 1997. Influence of oilseeds in combination with vitamin E supplementation in the diet on backfat quality of pigs. Anim. Feed Sci. Tech. 64: 91-100

Kouba, M. & J. Mourot, 1997. Effect of a high linoleic acid diet on Δ9-desaturase activity, lipogenesis and lipid composition of pig subcutaneous adipose tissue. Reprod. Nutr. Dev. 38: 31-37

Scheeder, M.R.L., H. Bossi & C. Wenk, 1999. Kritische Betrachtungen zur Fettzahl-Bestimmung. Agrarforschung 6: S1-8.

Suomi, K.. T. Alaviuhkola, J. Valaja, V. Kankare & A. Kempinnen, 1993. Effects of milk fat, hydrogenated vegetable oils on fat metabolism of growing pigs I. Growth, feed utilization and carcass quality in pigs fed different fats and oils. Agric. Sci. Finl. 2: 7-12

We are indebted to the Swiss Pig Performance Testing Station (MLP) for the very fruitful co-operation and to Centravo AG and the Commission for Technology and Innovation (KTI) for financial support.

Meat quality with reference to EUROP carcass grading system

W. Kapelanski, B. Rak, J. Kapelanska & H. Zurawski

University of Technology and Agriculture, Mazowiecka 28, 85-084 Bydgoszcz, Poland

Summary

The relationship between meat quality traits and carcass lean content was checked in 63 pigs graded according to EUROP classification. In each of the E, U, R, O classes there were 12, 13, 27 and 11 pigs, respectively. They were slaughtered at 105 kilograms liveweight. Carcass lean content (CLC) was predicted by UFOM-100 device. The CLC in each of the E, U, R, O classes attained 58.5 ± 3.2, 52.4 ± 1.4, 47.9 ± 1.3 and $42.2 \pm 1.2\%$, respectively. Meat quality generally tended to increase with the decrease of CLC. The pH_1 has increased from 5.97 in E to 6.49 in O class ($P < 0.01$). The LF_1 values were highest in E class 7.67 mS and gradually decreased to 3.36 mS in O class ($P < 0.01$). Drip loss was highest in E and lowest in O class ($P < 0.01$). The inverse was the case in water holding capacity of meat ($P < 0.01$). It is concluded that some meat quality deterioration may be expected in heavy muscled pigs.

Keywords: Pigs, Carcass grading, Meat quality

Introduction

There are many opinions that in pigs negative relationships exist between high muscularity and inadequate resistance to stress which in consequence leads to meat quality deterioration and PSE or DFD meat defect occurrence (Charpentier *et al.*, 1971; Rempel *et al.*, 1995; Sellier & Monin, 1994; Steinhauf *et al.*, 1976). That relationship is more marked in the most meaty pigs such as the Piétrain breed than in medium muscled pigs (Kapelanski *et al.*, 1999). The breed variation in meat quality can be ascribed to large breed differences in the RYR1 genotype frequencies in pig populations (Gronek, 1998; Sellier, 1998; Sosnicki *et al.*, 1998).

It will be stressed that, at present, the commercial pig producers are encouraged by the meat industry to produce the more meaty slaughter pigs due to the financial preference for high grade carcasses. In the EU countries the preferred model is that slaughter pigs ought to contain more than 55% of lean meat in the carcass. We can not exclude, therefore, that in future the production of heavy muscled pigs with the inferior meat quality will be preferred.

The aim of this study was to compare the meat quality traits in pigs classified into one of particular EUROP classes and to prove the relationship between quantity and quality of meat.

Material and methods

The study comprised of 63 crossbred pigs originating from the TORHYB crossbred programme (Kapelanski *et al.*, 1998). The pigs were the progeny of Piétrain boars mated to F_1 sows (Polish Large White x Polish Landrace). Growth performance was tested from about 25 kilograms until they attained 105 kilograms liveweight. The animals were fed *ad libitum* on complete feed mixture (12.8 MJ of ME, 175g crude protein and 11g lysine per 1 kilogram dry matter). Pigs were reared in groups of 10 pigs per pen. Upon reaching an individual liveweight of 105 kilograms they were transported to the slaughterhouse 22 kilometres away.

The animals were slaughtered in accordance with current regulations directly after arrival.

Carcass lean content was measured on warm left half-carcass by the ULTRA-FOM 100 device. After chilling the carcass measurements were done and valuable cuts were dissected in detail.

The pH_1 value of *longissimus lumborum* muscle was measured with the use of pistol pH-meter (Matthaus, Germany) and the meat electric concuctivity (Leitfähigkeitmessung) with the use of LF-Star apparatus (Matthaus, Germany). Drip loss was determined according to Honikel (1987), water holding capacity according to the Grau and Hamm method and was expressed as a percent of bound water, colour saturation and lightness as described by Rozyczka *et al.*, (1968). The fat and protein contents were assayed using commonly applied standard methods (AOAC, 1960).

The results obtained were statistically computed and variance analysis was performed according to computer programme Statistica (1995).

Results and discussion

The mean values for lean content and lean proportion in valuable cuts, as well as the loin eye area, average backfat thickness and total lean content in the carcasses graded according EUROP system are presented in Table 1.

Table 1. Carcass characteristics in relation to EUROP classes.

Traits	Classes				Differences	
	E	U	R	O	P<0.05	P<0.01
Number, n	12	13	27	11		
Ham lean content, kg	6.85	6.38	6.02	5.71	E - U	E-R,O
	0.76	0.38	0.55	0.53		U-O
Ham lean content, %	77.02	73.28	70.95	67.82	U - R	E-U,R,O
	2.20	1.90	3.61	3.05		O-U,R
Ham external fat, kg	1.44	1.71	1.86	2.13	NS	E-U,R,O
	0.15	0.16	0.29	0.20		O-U,R
Loin eye area, cm^2	49.53	47.48	42.80	40.96	U - R	E-R,O
	6.14	4.37	5.75	5.87		U-O
Loin lean content, kg	4.77	4.69	4.42	4.36	E - O	NS
	0.59	0.26	0.48	0.34		
Loin lean content, %	59.59	55.98	51.25	48.55	E - U	E-R,O
	2.87	2.62	4.85	2.80		U-R,O
Loin external fat, kg	1.89	2.37	2.86	3.31	NS	E-U,R,O
	0.23	0.24	0.54	0.30		U-R,O
						R-O
Average backfat thickness, mm	24.00	29.00	29.70	36.10	NS	E-U,R,O
	2.70	4.40	4.80	4.50		O-U,R
Carcass lean content, %	58.50	52.40	47.90	42.20	NS	E-U,R,O
	3.20	1.40	1.30	1.20		U-R,O
						R-O

NS - Non significant

It should be stressed that the difference in overall lean content or lean proportion in some cuts of particular carcass classes were somewhat lower than the differences in fat tissue content in the same cuts. It was more distinct in the loin cuts where the absolute lean content did not differ from E to O class, whereas the proportion of lean content decreased from 59.59% in E to 48.55% in O classes, due to the progressive increase of subcutaneous fat tissue from 1.89 kilograms in E to 3.31 kilograms in O classes. Generally, the lean content in carcasses graded into particular EURO classes was significantly different from 58.5 to 42.2% (P < 0.01).

Meat quality traits are presented in Table 2. Significant differences in such muscle traits which are ascribed to PSE meat, i.e. low pH_1 value, high LF_1 value, high drip loss or poor water holding capacity of meat, were stated between the extremely muscled classes (P < 0.01) and were inferior in E class. Similarly, the lightest meat colour was in E class and the darkest in O class (P < 0.05). Muscle fat and protein content were not different in respect to carcass lean content. The above results lead to the conclusion that some deterioration of meat quality may be expected in the heavy muscled pigs.

Table 2. *Meat quality traits in respect to carcass EUROP classes mean values (x) and standard deviations (s).*

Traits	Classes				Differences	
	E	U	R	O	P<0.05	P<0.01
pH_1	5.97	6.12	6.21	6.49	U - O	E - O
	0.62	0.42	0.40	0.41		
LF_1, mS	7.67	4.44	4.05	3.36	NS	E-U,R,O
	3.89	1.30	1.78	0.57		
Drip loss, %	6.52	5.03	4.39	2.14	R -E,O	E,U-O
	2.64	3.22	2.59	1.33		
WHC, bound water, %	68.51	72.34	72.39	74.78	E - R	E - O
	4.74	3.64	3.35	3.81		
Colour saturation, %	22.88	21.00	20.45	21.42	E - R	NS
	3.75	2.38	2.69	2.66		
Colour lightness, %	25.87	23.94	24.07	22.68	E - O	NS
	3.99	3.39	3.51	4.42		
Fat content, %	0.87	0.81	1.17	0.92	NS	NS
	0.22	0.29	0.54	0.60		
Protein content, %	23.39	23.65	23.62	23.19	NS	NS
	0.79	0.52	0.80	0.55		

NS - Non significant

References

AOAC, 1960. Official Methods of Analysis 9th Ed. Association of Official Analytical Chemists

Charpentier, J., G. Monin & L. Ollivier, 1971. Correlations between carcass characteristics and meat quality in Large White pigs. Proc. 2nd Int. Symp. on Condition and meat quality of pigs. Zeist, Pudoc, Wageningen, 255-260

Gronek, P., 1998. An analysis of genotype frequencies and genetic distances between some swine breeds. Pol. J. Food Nutr. Sci. 7/48, No 4(S):145-148

Honikel, K.O., 1987. The water binding of meat. Fleischwirtschaft, 67:1098-1102

Kapelanski, W., M. Bocian, J. Kapelanska, A. Hammermeister & S. Grajewska, 1998. Meat quality of porkers with a share of Pietrain blood. Prac. Mat. Zoot. Zesz. Spec. No 8: 91-97

Kapelanski, W., J. Kuryl, M. Bocian & B. Rak, 1999. Effect of RYR1 gene on meat quality traits in Polish Landrace, Pietrain and Zlotniki Spotted pigs. Adv. Agric. Res. vol. VI, Fasc 2 (in press)

Rempel, W.E., Ming YuLu, J.R. Mickelson & C.F. Louis, 1995. The effect of skeletal muscle ryanodine receptor genotype on pig performance and carcass quality traits. Anim. Sci. 60: 249-257

Rozyczka, J., J. Kortz & S. Grajewska, 1968. A simplified method of the objective measurement of colour in fresh pork meat. Rocz. Nauk Rol. 90-B-3: 345-353

Sellier, P., 1998. Major genes and crossbreeding with reference to pork quality. Pol. J. Food Nutr. Sci. 7/48 No 4(S): 77-89

Sellier, P. & G. Monin. 1994. Genetics of pig meat quality: A review. J. Muscle Foods, 5:187-219

Sosnicki, A.S., E.R. Wilson, A. de Vries & W.F. Wojcikiewicz, 1998. Efficient production of high quality pork: Genetic, environmental and marketing factors. Pol. J. Food Nutr. Sci. 7/48, No 4(S): 53-69

Steinhauf, D., J.H. Weniger & H.P. Mader, 1976. Observations on the apparent antagonism between producing capacity and meat quality. In Meat animals growth and productivity. Eds. D. Lister, D.N. Rhodes, V.R. Fowler & M.F. Fowler. Plenum Press, New York, 373-385

Pig Production and pork quality

The correlations between the fattening and slaughter performance in pigs

A. Pietruszka, R. Czarnecki & E. Jacyno

Department of Pig Breeding, Agricultural University of Szczecin, Dr. Judyma 10 Street, 71-460 Szczecin, Poland.

Summary

The relationship between the fattening performance and slaughter traits of carcass was determined on 120 fatteners, progeny of Polish Large White x Polish Landrace sows and boars of Piétrain , Piétrain x Polish Synthetic Line 990 and Piétrain x Duroc. The hybrid boars were made using the reciprocal cross. Together there were five groups. There were 24 pigs of each genotype (12 gilts and 12 barrows) in the animal experimental group. The animal were fattened from 34 to 100 kilograms of liveweight and slaughtered for testing. The daily body weight gains were correlated positively and highly significant with the percentage meat in carcass measured on live animals, but positively and nonsignificantly with the percentage meat in carcass measured after slaughter. Feed conversion per 1 kilogram daily body weight gain and leanness carcass was measured on live animals as well as after slaughter, and the correlation coefficients were negative and significant. Higher but still negative correlations were obtained between feed conversion per 1 kilogram meat gain and both measurements of percentage meat in carcass.

Keywords: Fatteners, Growth rate, Feed conversion, Carcass meatiness, Correlation coefficient

Introduction

Views on the correlation between fattening and slaughter performance traits have not always been univocal. This kind of view refers mostly to studies performed on the material characterised by lower meatiness. Selection carried out intensively for the improvement of meatiness in pigs had caused the opinion that the relationship between the fattening and the slaughter value traits have become clearer since it appeared that in more meaty pigs a positive relationship exists between the growth rate and the meatiness of the carcass (Jacyno & Pietruszka, 1997; Demo *et al.*, 1993; Michalska *et al.*, 1993; Czarnecki *et al.*, 1996). Moreover, pigs of a higher genetic potential for daily accumulation of protein, and thus for higher meatiness, better utilise protein and energy (Jacyno *et al.*,1996) as well as feed (Chabiera *et al.*, 1994; Demo *et al.*, 1993; Henkel, 1985; Hennebach *et al.*, 1987; Stiewe, 1986; Tribler, 1986) per 1 kilogram gain body weight, and thus the increase of fatless mass, compared with those characterised by lower meatiness. Therefore, in fatteners coming from different variants of commercial cross-breeding involving boars of "meaty" breeds positive correlations are found between the feed consumption and the backfat thickness whereas negative ones are found between the former and the eye muscle area, the weight of hind ham or the percentage of meat in carcass (Jacyno & Pietruszka, 1997; Czarnecki *et al.*, 1996; Buczyński *et al.*, 1998). In national cross-breeding programmes pigs of line 990 are used as a paternal component to produce slaughter material with the aim of ensuring in the first place the high meatiness. The results of studies carried out in the recent years by Duniec *et al.*

213

(1996) and Czarnecki *et al.* (1996), in which this line was used, have also confirmed that more meaty pigs are characterised by higher growth rate and lower feed consumption to achieve that increase.

Material and methods

The studies were carried out on progeny of Polish Large White x Polish Landrace sows and boars of Piétrain (P), Piétrain x Polish Synthetic Line 990 (P x L 990), L 990 x P, P x Duroc (D), D x P. The hybrid boars were made using the reciprocal cross. The experimental parts of this study were performed on 120 fatteners divided up into 5 genotype groups with 24 pigs in each (12 gilts and 12 barrows). The fattening started with animals at 34 kilograms of body weight and continued up to 100 kilograms. Within the fattening period the animals were maintained and fed on an individual basis with full-ration feed mix calculated according to the Nutrient Requirements of Pigs (1993). A mixture of 1 kilogram contained: 13,1 MJEM, 162 g CP, 8,6 g Lys, 5,8 Met + Cys. Upon the termination of the fattening period the animals were slaughtered and the basic cuts were dissected from the right half-carcasses according to the procedure applied at slaughter performance testing stations (Różycki & Rab, 1994). Moreover, immediately before slaughter of the pigs the meatiness was determined employing an ultrasonic apparatus PIGLOG 105 (with original equation of regression installed by the Denmark producer). For evaluation of feed conversion per 1 kilogram meat gain, the determination of the amount accumulated in the meat tissue was necessary. Therefore from the weight of the meat in basic cuts and the effective percentage meat content in the carcass, the amount of accumulated meat was estimated. The results of the research were analysed statistically applying a computer program "Statistica".

Results

Table 1.   Characteristic of the experimental material.

Traits	Average	Standard deviation
Daily body weight gain (g)	785	8,49
Intake of food: -per 1 kg body weight gain	2,94	0,30
-per 1 kg meat gain	4,67	0,16
Meat in carcass % on live pigs	51,8	3,09
Meat in carcass % after slaughter	53,6	3,28
Average backfat thickness with 5 measurements (cm)	2,38	0,09

The obtained results of growth rate and feed consumption per 1 kilogram gain in body weight for pig fattening in Poland can be admitted to be good. Slightly higher growth rate, from 816 to 830 g daily, is found in Yorkshire, Landrace, and L-990 purebred control fatteners evaluated in Polish Pig Slaughter Testing Stations (Różycki & Tyra, 1998). The analysed material was of slightly lower meatiness (2.38 cm) compared to all pig breeds evaluated in the Pig Slaughter Testing Stations (1.20 to 1.89 cm). It was also characterised by lower meat content in carcass (53.6%) compared to purebred material evaluated in the Pig Slaughter Testing Stations (from 65.86% for Piétrain to 54.43% for Yorkshire). It can also be noticed that the live-determined percentage of meat (51.8%) is lower when compared with that determined after slaughter and dissection (53.6%). This may result from the fact that the original regression equation used by the PIGLOG 105 apparatus is adapted to Danish material that is characterised by much higher meatiness compared to material analysed in the present work.

The correlation coefficients presented in Table 2 point to a positive relationship between the daily gains of body weight and the meat percentage in carcass. At the same time, a higher and highly significant coefficient of correlation (r=0.23**) was obtained between the gains and the percentage of meat determined with an ultrasonic meter before slaughter than between the former and the percentage of meat determined after slaughter (r=0.14). A low and non-significant correlation (r=-0.10) was obtained between the backfat thickness and the daily gain. This speaks, however, for a positive correlation between the meatiness of pigs and their growth rate mentioned in the beginning. This can be confirmed by a low though still positive correlation (r=0.16) between the feed consumption per 1 kilogram of gain and the backfat thickness, and a much higher, highly significant correlation (r=0.34**) between the backfat thickness and the feed consumption per 1 kilogram of meat gain. It should be emphasised that much higher correlation coefficients were obtained between the feed consumption per 1 kilogram meat gain and the studied meatiness traits of fatteners (-0.40**, -0.52**, -0.34**, respectively) compared to values for these correlations between the feed consumption per 1 kilogram body weight of fatteners and the same meatiness traits (-0.18*, 0.17*, 0.16, respectively; see Table 2).

Despite the fact that in the present study fatteners of lower fattening and slaughter value were analysed, the obtained relationships pointed to a positive correlation between these traits.

Table 2. Correlation coefficients between fattening and slaughter traits.

Traits	Daily body weight gain	Intake of food	
		per 1kg body weight gain	per 1kg meat gain
Carcass meat percentage (before slaughter, ultrasonic measurements)	0,23**	-0,18*	-0,40**
Carcass meat percentage (after slaughter, dissection)	0,14	-0,17*	-0,52**
Average backfat thickness with 5 measurements	-0,10	0,16	0,34**

Correlation coefficients significant at: *P≤0,05; **P≤0,01.

References

Buczyński, J.T., E. Fajfer & K. Szulc, 1998. Heritability and phenotypic and genetic correlations between certain fattening and slaughter traits in Polish Large White and Polish Landrace pigs. Pr. Mat. Zoot. 8: 105-112

Chabiera, K., M. Kotarbińska, S. Raj, H. Fandrejewski & D. Weremko, 1994. Effect of intake of metabolizable energy and lysine on the performance and chemical body composition of growing pigs. Mat. Konf. Nauk. Wpółczesne zasady żywienia świń. Jabłonna, 30-31 maj: 38-41

Czarnecki, R., K. Dziadek, M. Różycki, J. Owsianny & M. Kamyczek, 1996. The correlation between fattening and fleshing parameters of performance young boars and gilts 990 line. Mat. Konf. Nauk. Zootechniczno – ekonomiczne uwarunkowania mięsności świń. Rzeszów, 3-4 grudzień: 42-49

Demo, P., M. Letkovičova & L. Hetény, 1993. Analýza vztahov medzi ukazovateĺni výkrmnosti, jatočnej hodnoty a kvality mäsa hybridných ośipaných. Živoč. Výr. 38: 21-30

Duniec, H., M. Różycki & A. Szewczyk, 1996. Genetic, environment and phenotypic parameters some traits of 990 line. pigs Rocz. Nauk. Zoot. 23: 11-22

Henkel, D., 1985. Futteraufwand-ein Merkmal. Die Mühle und Mischfuttertechnik. 122: 325-328

Hennebach, H., G. Lengerken & H. Pfeiffer, 1987. Untersuchungen zur Futteraufnahme bei Mastschweinen. Tierzucht. 2: 77-79

Jacyno, E., R. Czarnecki, J. Dvořák, A. Pietruszka & B. Delikator, 1996. The correlation between fattening performance of parameters and fleshing characteristic in pigs. Mat. Konf. Nauk. Zootechniczno-ekonomiczne uwarunkowania mięsności świń. Rzeszów, 3-4 grudzień: 147

Jacyno, E., R. Czarnecki, J. Owsianny, K. Lachowicz & L. Gajowiecki, 1996. Fattening and slaughter values of pigs in relation to their genotype and protein source of their feeding diet. Arch. Anim. Nutr. 49: 169-179

Jacyno, E. & A. Pietruszka, 1997. The relationship between the fattening and slaughter performance in pigs. Advances in Agricultural Sciences. Szczecin, 34: 47-51

Michalska, G., J. Nowachowicz, B. Rak, J. Kapelańska & W. Kapelański, 1993. The relationship between the fattening performance and the carcass traits in Duroc pigs. Zesz. Nauk. Przegl. Hod. 9: 149-153

Nutrient Requirements of Pigs, 1993. (tables), Inst. of Animal Physiology and Nutrition, Jabłonna, Omnipres: 1-87

Różycki, M. & K. Rab, 1994. Results of pigs tested at pig testing stations, Report on pig breeding in Poland. Inst. Zoot. Kraków: 66-93

Stiewe, H., 1986. Futteraufnahme und Futterverwertung. Die richtigen Tiere zur Mast einstellen. Schweinez. u. Schweinemast. 34: 308-310

Tribler, G., 1986. Züchterische Aspekte der Futterverwertung beim Schwein. Z. der Humboldt-Universität Berlin, Math.-Nat. R. 35 (4).

Comparison of fat supplements of different fatty acid profile with growing-finishing swine

J. Gundel*[1], A. Hermán Ms.[1], M. Szelényi Ms[1], & G. Agárdi[2]

[1]*Research Institute for Animal Breeding and Nutrition, 2053 Herceghalom, Hungary,*
[2]*Debrecen University of Agricultural Sciences, P.O Box. 36, 4015 Debrecen, Hungary*

Summary

Hungarian LW X Dutch LR (n=4x12, kept in individual pens keeping, 30.5 kg BW), females and barrows were fed with four types of corn – barley based isocaloric, isoproteic, iso-AA, and iso-lipide rations (ME:13.3 MJ/kg, CP: 16.3 %, LYS: 0.95 %, EE: 4.3 %), using 1) animal fat (AF), 2) fullfat sunflower seed (FFSF), 3) sunflower expeller cake (SFEC), and 4) fullfat soybean meal (FFSB) as fat sources of different linoleic acid levels. The linoleic acid levels (%) in the diets were: 1. (AF): 1.5, 2. (FFSF): 2.0, 3. (SFEC): 2.54, (FFSB): 2.1.

Slaughtered at a liveweight of 98.1 kilograms, the following performances were obtained (in the same order as the diets): daily gain: 624, 655, 657, 690 g; feed conversion; 3.55, 3.21, 3.30, 3.14 kg/kg; lean meat, %: 52.1, 52.4, 52.9, 53.5; backfat, mm: 25.4, 24.5, 25.2 and 24.3.

Comparing fattening performances and carcass quality the vegetable fat sources (particularly fullfat soybean) proved to be superior to animal fats. Vegetable fats increased the linoleic acid in the carcass fat related to animal fats (100%) in the following order: FFSF: 168 %, SFEC: 147%, FFSB: 131 %.

Keywords: Growing-finishing pig, Sunflower seed, Sunflower cake, Soybean seed, Unsaturated fatty acids

Introduction

In Hungarian pig feeding the source of protein, as in many other countries, is imported extracted soybean and there is a permanent aim to replace it with a home product. There is another aim too, namely to fulfil the demand of the consumers who need less fat with a specific fatty acid profile in the pork meet. With our experiment we tried to answer both questions in Hungarian conditions. We used as source protein and unsaturated fatty acids extruded fullfat soybean, fullfat sunflower, sunflower cake (cold pressed) and as a control extracted sunflower + fat (animal origin). On the other hand using these feeds can realise a so called "high density" feeding of growing-finishing pigs.

A common characteristic of these three feeds is the high fat(oil), protein and energy content. Using these feeds in a suitable composition could be very advantageous, among other things they promote the deposition of protein, give assistance to the ossification, increase the absorption of the fat soluble vitamins, and mainly they modify the fatty acid composition of the fat-tissue, and generally improve the taste of feed as has been published by many researchers (Kállai, 1953; Berschauer *et al.*, 1983, 1984; Fekete, 1995; Schmidt, 1996). Madsen *et al.* (1992) summarised in their review the Danish experiences of the last 40 years concluding that the relationship between the amount and composition of dietary fat and the content and profile of intramuscular fat seems to be complicated. In their opinion oats, rapeseed, sunflower seed, animal fat and vegetable oil should only be included in limited amounts in the diet to avoid high occurrence of unacceptable soft backfat and rancidity of the carcass.

The aim of our experiment was to determine the effect of the concentrates with same nutrient content but different fatty acid amount and composition on the result of the fattening, and the fatty acid composition of backfat.

Materials and methods

Hungarian LW X Dutch LR (n=4x12, kept in individual pens, initial BW 30.5 kg), females and barrows were *ad lib* fed individually with four types of corn – barley based isocaloric, isoproteic, iso-AA, and iso-lipide rations (ME:13.3 MJ/kg, CP: 16.3 %, LYS: 0.95 %, EE: 4.3 %), using 1) animal fat (AF), 2) fullfat sunflower seed (FFSF), 3) sunflower expeller cake (SFEC), and 4) fullfat soybean meal (FFSB) as fat sources of different linoleic acid levels. The linoleic acid levels (%) in the diets were: 1. (AF): 1.5, 2. (FFSF): 2,0, 3. (SFEC): 2.54, (FFSB): 2.1. (*Table 1.*). The slaughtered pigs (mean BW was 98 kg) were qualified by EUROP system and the fatty acid composition of the back fat samples was determined by the GC method.

Table 1. The composition of concentrates (%).

Treatments	AF	FFSF	SFEC	FFSB
Maize	37,00	37,00	37,00	37,00
Barley	34,81	36,92	36,45	38,95
E. soybean, 46 %	9,40	18,00	15,00	12,00
Extracted sunflower, 40 %	12,00	—	—	—
Fullfat sunflower	—	5,50	—	—
Sunflower cake (cold pressed)	—	—	9,00	—
Fullfat soybean (extruded)	—	—	—	9,43
Fat powder-40	4,00	—	—	—
Chalk	1,05	1,00	1,00	1,00
MCP	0,50	0,50	0,50	0,55
NaCl	0,30	0,30	0,30	0,30
L-lizin-HCl	0,34	0,18	0,18	0,17
Kensym Dry HF (high fibre)	0,10	0,10	0,10	0,10
Premix, 0.5 %	0,50	0,50	0,50	0,50

Chemical composition: DM 86,7; DEs 13,9 MJ/kg; MEs 13,3 MJ/kg; CP 16,3 %; EE 4,3 %; CF 4,6-5,1-5,8-3,9 %; LYS 0,95 %; MET+CYS 0,57 %; Ca 0,66 %; P 0,52;

Results and conclusion

The results of the experiment can be summarised as follows (*Table 2.*): the most constant daily gain was shown by the treatment "SFEC", but the best was the "FFSB". This treatment was better than "AF" by 10%, "FFSF" by 5.2% and "SFEC" by 4.8% (all differences are P<5%). The best feed conversion ratio (3.14 kg/kg) was in the "FFSB" treatment too, while the worst was in "AF" (P<5%). Averages of the daily feed intake (2.10-2.21kg, NS) were not influenced by the treatments.

Table 2. Fattening performances in the experiment.

Treatments	AF		FFSF		SFEC		FFSB	
	Mean	s	Mean	s	Mean	s	Mean	s
Initial weight, kg	30,4	4,4	31,5	4,4	29,4	4,1	29,2	5,0
Final weight, kg	97,2	2,7	99,0	3,2	98,0	3,5	98,2	2,8
ADG, g	624	41,0	655	39,0	657	58,7	690	43,7
Feed intake, kg/day	2,21	0,1	2,10	0,2	2,17	0,2	2,17	0,1
FC, kg/kg	3,55	0,2	3,21	0,3	3,30	0,2	3,14	0,2

On the other hand it seems, that in the growing period the fullfat sunflower can be used less (maybe it has too much crude fibre) than in the finishing period in which better gain was produced by the "AF" (extr.sunflower+fat animal origin) treatment.

The result of the slaughter qualification is summarised in *Table 3*. We found the highest lean meat content in the treatment "FFSB" (53.5%) and the lowest in the "AF" ($p < 5\%$). The tendency was the same for the backfat thickness ("FFSB" 24.3 mm, "AF" 25.4 mm) but the differences were not significant. It is in agreement with the experiences of Marcello *et al.* (1984) who found that feeds with high oil content could not be fed to a pig unrestrictedly, for example more than 13 % fullfat sunflower seed decreases the slaughter quality.

Table 3. Slaughter qualification.

Treatments	FA		FFSF		SFEC		FFSB	
	Mean	s	Mean	s	Mean	s	Mean	s
Slaughter weight, kg	97,2	2,7	99,0	3,2	98,0	3,5	98,2	2,8
Backfat thickness, mm								
at withers	35,1	5,8	33,1	3,1	33,7	5,4	34,3	4,2
at middle of back	22,4	3,6	21,4	3,9	22,1	4,5	20,7	2,6
at loins	18,8	1,9	19,1	3,5	19,7	4,8	17,8	3,4
average	25,4		24,5		25,2	7,8	24,3	
Lean meat, %	52,1	1,8	52,4	2,2	52,9	3,9	53,5	2,4

Contents of the unsaturated fatty acids of the concentrate and of the samples of the backfat are presented in *Table 4*. We have found that the fatty acid composition of the concentrate is reflected in the backfat samples. It confirms the opinion of Courbourlay & Massabie (1994) who found that 4 % fullfat sunflower can modify the composition of backfat (e.g. the amount of the $C18:2$ in relation to the control increased from 4.4 to 21.0%). In our experiment a particularly good example is the "FFSB" treatment, because in soybean there is linolenic acid ($C18:3$) and in the sunflower there is not and accordingly we found it only in samples of the "FFSB" backfat.

As a conclusion we could have found that the different fatty acid compositions did not influence significantly the daily feed intake, but from the other point of the fattening performance the "FFSB" (extruded fullfat soybean) was the best, as we know from our other experiment, probably because the digestibility of the nutrients are better in soybean than in the sunflower. The sunflower treatments had given good results too and using it depends on the price situation. It seems also that the fatty acid composition of the concentrates can influence the fatty acid composition of the backfat.

Table 4.   Unsaturated fatty acid composition of the concentrates and the backfat.

Treatments	In the concentrates				In the backfat			
	AF	FFSF	SFEC	FFSB	AF	FFSF	SFEC	FFSB
oil acid (C18:1)	34.8	25.5	23.9	22.3	43.6	37.8	38.6	39.9
linoleic acid (C18:2)	28.8	56.7	55.7	53.4	13.6	23.0	20.1	18.7
linolenic acid (C18:3)	1,0	1.4	1.4	4.3	0	0	0	0,9
other unsaturated acids	3.2	1.6	2.7	2.1	3.1	2,7	2.9	2.8

In our opinion further investigations are needed because consumers continue to demand less fat with a specific fatty acid profile, and these investigations must be concentrated on utilising the possibilities of manipulating the fat content and fatty acid composition of individual fat depots. It has been shown that changes can occur as a consequence of dietary variations, but the physiological upper and lower limits of fat amounts and fatty acid composition must also be established from an animal welfare point of view. Further researches are needed on regulatory factors of the fat (fatty acid) metabolism.

References

Bersauer, F., J. Rupp & U. Ehrensvärd, 1984. Untersuchungen über ernährungsphysiologische Wirkungen von Futterfetten in Rationen für wachsende Schweine. 2. Mitt.: Einfluss von Sonnenblumen- und Kokoskernen auf den Protein- und Fettansatz, das Fettsäurenmuster des Rückenspecks sowie auf einige Blutparameter. Arch. Tierernährung, 34: 19-23

Bersauer, F., U. Ehrensärd & K.H. Menke, 1983. Untersuchungen über ernährungsphysiologische Wirkungen von Futterfetten in Rationen für wachsende Schweine. 4. Mitt.: Einfluss von Sonnenblumenöl und Kokosfett auf den Protein- und Fettansatz, das Fettsäurenmuster des Rückenspecks sowie auf einige Blutparameter. Arch. Tierernährung, 33: 826-842

Courboulay, V. & P. Massabie, 1994. Use of sunflower seed in diets for growing pigs: effect on performance and backfat quality. Journées Rech. Porcine en France, 26: 207-211

Fekete, L., 1995. Sertéstakarmányozás. Mezőgazda Kiadó, Budapest, Hungary, 292pp

Kállai, L., 1953. Adatok a napraforgóolajok biológiai értékéhez és változásához. Állattenyésztés, 2: 151-160

Kurelecz, V., 1950. Napraforgó-pogácsa és extrahált napraforgó-dara használhatósága a malacnevelésben. Agrártudomány, 2: 370-374

Madsen, A., K. Jacobsen & H.P. Mortensen, 1992. Influence of dietary fat on carcass fat quality in pigs. A review. Acta Agric. Scand., Sect. A, Animal Sci. 42: 220-225

Marchello, M. J., N.K. Cook, V.K. Johnson, W.D. Slanger, D.K. Cook & W.E. Dinusson, 1984. Carcass quality, digestibility and feedlot performance of swine fed various level of sunflower seed. J. Anim. Sci. 58: 1205-1210

Schmidt, J., 1996. Takarmányozástan. Mezőgazda Kiadó, Budapest, Hungary, 358pp

The pork meat quality in pigs with a different intensity of nitrogen substances retention

M. Čechová[1], V. Prokop[2], K. Dřímalová[1], Z. Tvrdoň[1] & V. Mikule[1]

[1] *Department of Animal Breeding, Mendel University of Agriculture and Forestry, Zemědělská 1, Brno 613 00, Czech Republic*
[2] *Research Institute for Animal Nutrition, Ltd., Pohořelice 691 23, Czech Republic*

Summary

In final carcass pig hybrids interbreeding combination No.1(White Improved x Landrace) x Large White and No. 2 (White Improved x Landrace) x (Czech Meat Pig x Piétrain) in a balanced experiment for a weight of 48 kilograms, limits were found for nitrogen deposition in pig combination No.1 at a level of 147 g per pig per day and in pig combination No. 2 at a level of 165 g per pig per day. After fattening until a weight of 100 kilograms (pigs were fed with the same feed mixture) some meat quality traits were determined in the eye muscle: combination No.1 - pH_1 – 6,40, dripping water losses 4,11 and content of intramuscular fat 3,25, combination No.2 - pH_1 – 6,40, dripping water losses 4,35 and content of intramuscular fat 2,53. These results weren't statistically conclusive.

Keywords: Nitrogen deposition, Dripping water losses, Intramuscular fat, pH_1

Introduction

There are different ways to develop breeding and fattening of pigs - genetic and environmental factors. The animal's genotype needs special environmental conditions for optimal manifestation of genetically fixed abilities. Nutrition is a prior factor, and there is an important back connection to genotype (genotype determines genetic fixed abilities of nutrients and energy deposition - especially as proteins and fat).

Determination of these abilities in relatively homogenic types of the most often used hybrid combination of growing pigs is important for effective breeding and improving work and for determination of nutrition conditions.

An identification of genetically fixed abilities of protein deposition in a growing pig body of two hybrid combinations (WI x L) x (CMP x Pn), (WI x L) x LW and monitoring of chosen traits of meat quality were the aim of the study. WI - White Improved, CMP - Czech Meaty Pig.

Material and Methods

We carried out two balance tests on growing pigs of two defined hybrid combinations (WI x L) x (CMP x Pn) and (WI x L) x LW.

There were 8 hogs (half sibs – one father, different mothers) in each balance test, the test ran for an average weight of 45-48 kilograms.

There were four balances in each test, each test's intervention was attested by all of the eight pigs by a latin square method. The same feed mixtures were used during study.

Total nitrogen balance was observed. Tests ran in four groups with four feeding mixtures with different levels of nitrogen substances and aminoacids (screening).

After the end of the balance tests pigs were used for finishing test until the weight of 100 kilograms. Then they were slaughtered and samples from an eye muscle were taken. We determined traits – pH_1, dropping water loses and content of intramuscular fat.

Conclusions and Discussion

In the first test there were 8 hogs of the combination (WI x L) x LW. A specific fixed limit for deposition of nitrogen substances was determined on a level 147 g of nitrogen substances per day. In the second test, there were 8 hogs of the hybrid combination (WI x L) x (CMP x Pn) tested. A specific fixed limit for deposition of nitrogen substances was higher on a level 165 g of nitrogen substances per day (Table 1).

Table 1. Deposition of nitrogen substances.

Combination	A limit of NS deposition (pig/day in g)
(WI x L) x LW	147
(WI x L) x (CMP x Pn)	165

Campbell & Taverner (1985) wrote, that pigs of top meat hybrids are able to deposit 200 g NS per day. During the following analysis and determination of some meat quality traits we determined the same level of $pH_1 = 6,4$. Quite similar were dripping losses of water (4,11 % and 4,35 %).

In the combination (WI x L) x LW a higher content of intramuscular fat was determined – 3,25 %, in the combination (WI x L) x (CMP x Pn) – 2,53 %.

Differences among traits values weren't statistical conclusive. In comparison with the data of Nürnberg & Ender (1990), Glodek (1990), our levels are higher and determine a high pork quality of tested hybrids.

Table 2. Results of meat quality analyses.

Trait	n	average	s_x	n	average	s_x
pH_1	4	6,40	0,08165	4	6,40	0,08165
water losses	4	4,35	1,29092	4	4,11	0,74006
IM fat	4	2,53	0,74677	4	3,25	1,04933

References

Campbell, R.G. & M.R. Taverner, 1985. Effect of strain and sex on protein and energy metabolism in growing pigs, Energy Metabolism of Farm Animals, EAAP Publication 32: 78-81

Glodek, P., 1990. Was kommt nach der Züchtung auf Stressresistenz? SUS, 38: 300-303

Nürnberg, K. & K. Ender, 1990. Aktuelle Ergebnisse zur Fettsäurezusammensetzung und Fettqualität, Schweinefleischbeschaffenheitnach der Halothansanierung, Nordhausen, 210-218

Influence of nutrition on pork quality

The effect of dietary Ca-fatty acid salts of linseed oil on cholesterol content in longissimus dorsi muscle of finishing pigs*

T. Barowicz

National Research Institute of Animal Production, 32-083 Balice, Poland

Summary

36 Polish Landrace fatteners of both sexes, divided into 3 groups, were fed from 70 to 100 kilograms liveweight with a complete mixture containing 0 (control), 8 or 15 percent of Ca salts of fatty acids (CaFAS) made partly from linseed oil and blended fat in the proportion 1:1. The mixture contained 83% UFA, including 36% PUFA, while the proportion of PUFA *n-6* to PUFA *n-3* was 1: 1.6. The experiment concluded with slaughter. After dissection, a sample of *longissimus dorsi* muscle was taken to determine total cholesterol after lipid extraction. Total cholesterol was assayed using the enzymatic method.

Total cholesterol in *longissimus dorsi* muscle of experimental fatteners was observed to drop from 55.70 to 52.49 mg/100 g fresh tissue. The differences were particularly noticeable in the meat of sows, amounting to 57.14 (control), 52.68 and 51.11 mg/100 g fresh tissue, respectively. However, the differences were not statistically significant.

It is suggested that CaFAS made partly from linseed oil, especially as a 15 percent supplement to complete mixture for finishing pigs, may be one of the ways of improving the dietary value of pork.

Keywords: Pigs, Feeding Ca salts of fatty acids, PUFA, Meat, Cholesterol

Introduction

The high consumption of pork and pork products, coupled with an increasing proportion of people afflicted with atherosclerosis and atherosclerosis-based diseases, make it necessary to look for ways of restricting carcass fatness, changing its fat composition, and limiting its cholesterol content, thus improving the dietetic value of pork (Goodnight, 1993). Dietary PUFA acids decrease the level of cholesterol and its low density lipoprotein fractions in blood serum (Nettleton, 1991; Ulbricht, 1992). Our studies failed to show a clear effect of linseed on cholesterol content of fattening pig tissues (Barowicz *et al.*, 1997). 4 and 8% supplements of full-fat linseeds decreased the cholesterol level in blood serum while having no significant influence on the cholesterol content of the *longissimus dorsi* and cardiac muscle. The problem of how to decrease the cholesterol content of meat by way of nutrition is still unresolved. Determining the optimum relationship between dietary PUFA *n-6* and PUFA *n-3* may prove helpful in resolving this problem. This is confirmed by a recent experiment with rats, where the cholesterol content of fat decreased by 30% when various proportions of *n-6* and *n-3* fatty acids were used (Barowicz, 1998).

The aim of the studies was to determine the usefulness in the finishing period of complete mixtures containing Ca salts of fatty acids made partly from linseed oil – with a narrowed ratio of PUFA *n-6* to PUFA *n-3* - on total cholesterol content in the *longissimus dorsi* muscle.

* The study was financed by the Committee for Scientific Research (grant No 5 PO6E 058 14)

Material and methods

The feeding trial was carried out at the Zootechnical Experimental Station belonging to the National Research Institute of Animal Production in Rymanów. 36 Polish Landrace fattening pigs of both sexes were given 0, 8 or 15% friable fodder fat in complete feeds from 70 to 100 kilograms body weight. The experiment ended with the slaughter of animals. After dissection the *longissimus dorsi* muscle samples were assayed for the total cholesterol content according to the method given by Rhee *et al.* (1982) by extracting fat from tissues using the Folch *et al.* (1957) method. The concentration of total cholesterol was assayed in these extracts using the enzymatic method. The results were analysed statistically using the Statgraphics Plus 3.3 package.

Results

Table 1.   Composition (%) , nutrient content and fatty acid composition (in % of total fats) of complete feeds.

Item	Group		
	control	feed fat	
		8%	15%
Ground barley	66.80	55.00	56.30
Ground wheat	15.00	16.00	5.00
Soybean meal	15.00	17.60	20.00
Meat-and-bone meal	0.40	0.60	1.00
Friable feed fat	-	8.00	15.40
Fodder salt	0.25	0.25	0.20
Dicalcium phosphate	0.30	0.30	0.30
Limestone	1.00	1.00	0.80
L-lys HCL	0.25	0.25	0.20
Premix PW-2	1.00	1.00	0.80
Total	100.00	100.00	100.00
1 kg mixture contained:			
crude protein (g)	152.40	152.00	152.00
digestible crude protein (g)	124.19	125.29	124.98
crude fat (g)	29.92	24.76	28.71
crude fibre (g)	48.59	45.15	44.61
ME (MJ/kg)	12.38	12.73	12.95
crude protein/1 MJ ME (g)	12.31	11.94	11.74
calcium (g)	5.70	6.44	6.64
phosphorus (g)	4.56	4.40	4.34
Lys (g)	9.11	8.79	8.45
Met + Cys (g)	5.18	5.07	4.93
Fatty acids:			
Unsaturated acids (UFA)	60.64	70.51	76.95
Monounsaturated acids (MUFA)	22.33	43.02	51.29
Polyunsaturated acids (PUFA)	38.31	27.49	25.66
Polyunsaturated acids *n-3* (PUFA *n-3*)	4.99	8.97	11.46
PUFA *n-6*/PUFA *n-3*	6.68	2.06	1.24

The composition, nutrient content and fatty acid composition of the experimental feeds are given in Table 1. Table 2 shows the composition of fatty acids in experimental fodder fat. The experimental mixtures were characterised by a much higher UFA (especially PUFA *n-3*) content. As a result, their PUFA *n-6* to PUFA *n-3* ratio decreased.

Table 2. *Fatty acid composition (in % of total fats) of feed fat (calcium salts of linseed oil fatty acids).*

Fatty acids	Feed fat
C 8	0.02
C 10	0.04
C 12	0.05
C 14	0.66
C 16	7.30
C 16 : 1	2.11
C 18	8.79
C 18 : 1	44.51
C 18 : 2 *n-6*	13.55
g C 18 : 3 *n-6*	0.03
C 18 : 3 *n-3*	21.68
C 20	0.11
C 20 : 4 *n-6*	0.31
C 20:5 *n-3* (EPA)	0.15
C 22	0.06
C 22 : 1	0.26
C 22:6 *n-3* (DHA)	0.37
Saturated acids (SFA)	17.03
Unsaturated acids (UFA)	82.97
Monounsaturated acids (MUFA)	46.88
Polyunsaturated acids (PUFA)	36.09
Polyunsaturated acids *n-3* (PUFA *n-3*)	22.20
Dietary fatty acids having desirable neutral or hypocholesterolemic effect in humans (DFA)*	91.75
Dietary fatty acids having undesirable (hypercholesterolemic) effect in humans (OFA)*	8.25
PUFA *n-6* / PUFA *n-3*	0.63

* DFA - total unsaturated fatty acids (UFA) + C 18
* OFA - C 14 + C 16

Table 3. *Total cholesterol content (mg/100 g fresh tissue) in longissimus dorsi muscle of fatteners.*

Item	Group control	feed fat 8%	feed fat 15%	SEM*
Gilts (n=6)	57.14	52.68	51.11	1.19
Barrows (n=6)	54.26	53.57	53.87	1.32
Total (n=12)	55.70	53.12	52.49	0.87

* - standard error of the mean

Total cholesterol in *longissimus dorsi* muscle of experimental fatteners was observed to decrease from 55.70 to 52.4 9 mg/100 g fresh tissue (Table 3), being statistically non-significant. Particular differences were observed in the meat of gilts, where total cholesterol was 57.14 (control), 52.68 and 51.11 mg/100 g fresh tissue. These differences were not significant, either.

Conclusion

It is suggested that Ca-fatty acid salts made partly from linseed oil, especially as a 15 percent supplement to complete mixture for finishing pigs (from 70 to 100 kg b.w.) may be one of the ways of improving the dietary value of pork.

References

Barowicz, T. , F. Brzóska, M. Pietras & R. Gąsior, 1997. Hypocholesterolemic effect of full-fat flax seeds in the diets of growing pigs. Medycyna Wet. 53: 164-167

Barowicz, T., 1998. Dietary and tissue fat quality. Feed Mix, 6: 31-34

Folch, J., M. Lees & G.H. Sloane Stanley, 1957. A simple method for the isolation and purification of total lipids from animal tissue. J. Biol. Chem., 226: 497-509

Goodnight, S.H., 1993. The effects of n-3 fatty acids on atherosclerosis and the vascular response to injury. Arch. Pathol. Lab. Med. 117:102

Nettleton, J.A., 1991. Omega-3 fatty acids: Comparison of plant and seafood sources in human nutrition. J. Am. Diet Assoc. 91: 331-337

Rhee, K.S., T.R. Dutson & G.C. Smith, 1982. Effects of changes in intramuscular an subcutaneous fat levels on cholesterol content of raw and cooked beef steaks. J.Food Sci. 47: 1638-1642

Ulbricht, T.L.V., 1992. Animal fats and human health. Anim. Prod. 54: 462

Soybean oil, sex, slaughter weight, cross-breeding - influence on fattening performance and carcass traits of pigs

R. Kratz[1], E. Schulz[1], G. Flachowsky[1] & P. Glodek[2]

[1] *Federal Agricultural Research Centre (FAL) Braunschweig, Institute of Animal Nutrition, Bundesallee 50, 38116 Braunschweig; email: ruediger.kratz@fal.de*
[2] *Georg-August-University Göttingen, Institute of Animal Breeding and Genetics, Albrecht-Thaer-Weg 3, 37075 Göttingen*

Summary

In a feeding experiment with 96 pigs the influence of soybean oil (0% or 2.5 %), sex (barrows compared with gilts), slaughter weight (110 kg compared with 120 kg live weight) and sire line [Pi (NN) / Pi (PP) / Pi*Ha (NN) x LWxDL (NN)] on fattening performance and carcass traits have been determined. The pigs were fed individually and isoenergetically. Soybean oil had no effect on experimental parameters. Average daily gain and energy conversion rate of castrated males or gilts at 110 kilograms and 120 kilograms slaughter weight respectively were not significantly different. Gilts may reach 120 kilograms slaughter weight with the same fattening performance and greater ham and ham-lean weights than castrated males at 110 kilograms. Crosses did not differ significantly in average daily gain or energy conversion rate. Crosses with the homozygous MHS-positive Piétrain (Pi [PP]) boars had the highest ham and ham-lean weights, crosses with Pi*Ha the lowest weights and Pi (NN) were in the middle. At higher slaughter weight (110 kg to 120 kg) differences became smaller.

Keywords: Soybean oil, Sex, Slaughter weight, Crosses, Piétrain, Pig fattening, Meat percentage, Primal cuts

Introduction

High daily live weight with lean gain, with high carcass and meat quality should be obtained by pig fattening. To meet the demands of the meat industry variation in meat percentage, proportion of primal cuts and several meat quality characteristics should be small. Factors, such as genetics, feeding and management influence the uniformity of meat. A fattening trial was conducted to investigate how modern fattening practice has to look like to meet the demands of high quality pig meat. Results on fattening performance and carcass proportions shall be presented.

Material and Methods

The feeding trial was designed with 96 fattening pigs which were randomly assigned to the following factors:

Factor 1:   FEED   0 % and 2,5 % soybean oil,
Factor 2:   WEIGHT   110 kg and 120 kg slaughter weight,
Factor 3:   SIRE   Pi (NN), Pi (PP), Pi*Ha (NN),
Factor 4:   SEX   castrated males and gilts.

The pigs in the trial were crosses from the sires Pi (NN), Pi (PP), Pi*Ha (NN) with LWxDL (NN) dams. Grower (30 kg to 60 kg liveweight) and finisher (60 kg to 110 kg / 120 kg liveweight) diets consisted of barley, wheat, soybean meal, amino acids, vitamins, minerals and for group two 2.5 % soybean oil. Analysed composition and calculated energy content (GfE, 1987) are shown in Table 1. The pigs were kept in single boxes and fed isoenergetically. The pigs were weighed every week to formulate their daily food supply to meet an average daily gain of about 760 g/d.

Table 1. Nutrient composition [g/kg DM] and energy content [MJ ME/kg DM] of grower and finisher diets.

	grower diet		finisher diet	
soybean oil	-	2,5 %	-	2,5 %
XP	202	201	178	174
XL	22	49	22	50
XF	45	42	44	46
NfE	677	654	707	681
XA	54	54	49	49
Lysin	11	12	10	10
ME	14,8	15,5	14,8	15,3

The performance test started at an initial liveweight of 30 kilograms and finished on reaching final weight (110 kg and 120 kg respectively). Food consumption was determined daily. After chilling (24 h p.m.) carcasses were dressed using standard procedures (ALZ, 1997). Meat percentage was calculated according to the Bonner Formel equation. Ham weight included skin, fat and bones. Ham-lean weight was the primal lean meat from ham without skin, fat, bones and processing meat. Statistical analyses were directed with the procedure General Linear Models (GLM) of SAS/STAT® version 6.12. The hypotheses were tested with Duncan's test. Significant differences were determined if probabilities were ≤ 0.05.

Results

Fattening performance with average daily gain (ADG) of 820 g/d and energy conversion rate (ECR) of 35.5 MJ ME/kg indicated high performance. Soybean oil had no significant effect (ADG: + 16 g/d; ECR: - 0.6 MJ ME/kg ADG) on fattening performance. Fattening performance of gilts and castrated males at each weight class did not differ significantly in our study (Table 2). At rationed feeding gilts have higher daily liveweight gain than castrates, in contrast to *ad libitum* feeding. Therefore castrates reached higher ADG and food intake, which was determined in several studies (Ellis et al., 1996, Cisneros et al., 1996, Leach et al., 1996).

Table 2. Fattening performance of gilts and castrated males at 110 kg and 120 kg slaughter weight [n = 24].

WEIGHT	110 kg		120 kg	
SEX	gilt	castrate	gilt	castrate
ADG[1]	839[a]	823[ab]	819[ab]	800[b]
ECR[2]	33.5[c]	35.2[bc]	36.0[ab]	37.4[a]

[a, b, c] = means with different index letter per row are significantly different (p ≤ 0.05) [1] = average daily gain
[2] = energy conversion rate

Both sexes had better ECR (p ≤ 0.05) at 110 kilograms in comparison with 120 kilograms. Differences of ADG and ECR were significant for 110 kilogram gilts and 120 kilogram barrows. Gilts at 120 kilogram slaughter weight had a fattening performance like castrates at 110 kilograms, indicating that separate fattening of castrated males and gilts could be evident for economic pig production.

With increasing slaughter weight of 10 kilograms ADG decreased in tendency (- 22 g/d) and ECR increased significantly (2.3 MJ ME/kg ADG, p ≤ 0.05). To indicate fattening performance at several slaughter weights ECR is more precise. Under *ad libitum* feeding ADG was constant at 110 to 140 kilograms liveweight in the studies of Ellis & Avery (1990) and Kuhn *et al.* (1997), but food conversion rate and ECR raised significantly with increasing slaughter weight.

Sires did not differ significantly in fattening performance at each slaughter weight (Table 3).

Table 3. Fattening performance of sires at 110 kg and 120 kg slaughter weight [n = 16].

WEIGHT	110 kg			120 kg		
SIRE	Pi (NN)	Pi (PP)	Pi*Ha	Pi (NN)	Pi (PP)	Pi*Ha
ADG[1]	824	847	836	807	818	802
ECR[2]	34.8[abc]	33.9[c]	34.5[bc]	36.9[a]	36.2[ab]	37.0[a]

[a, b, c] = means with different index letter per row are significantly different (p ≤ 0.05)
[1] = average daily gain
[2] = energy conversion rate

Meat percentage of about 58.7 % (Bonner Formel; ALZ, 1997) indicated high performances of the used pig crosses. Soybean oil did not influence carcass traits of pigs. Gilts had about 3 % more lean meat than castrates (p ≤ 0.05). Ham-lean weights decreased slightly from 8.4 % of carcass weight at 110 kilograms to 8.2 % at 120 kilograms slaughter weight. Gilts had higher ham and ham-lean weights than castrated males at both slaughter weights (Figure 1). These differences were for the ham weights only at 120 kilograms not significant.

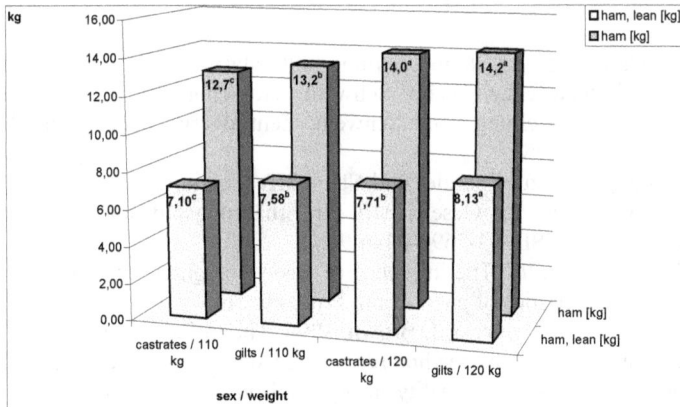

[a, b, c] = means with different index letter per row are significantly different (p ≤ 0.05)

Figure 1. Weights of ham and ham-lean from castrated male and gilts at 110 kg and 120 kg slaughter weight [n = 24].

At 110 kilograms slaughter weight Pi (PP) had the heaviest ham and ham-lean weights (p ≤ 0.05) (Figure 2). At 120 kilograms the hams of Pi (PP) and Pi*Ha had significantly different weights and Pi (NN) were in between. The ham-lean weights were not significantly different but in the same ranks. The differences in weights of ham and ham-lean became smaller with higher slaughter weight, according to Ellis *et al.* (1996), who found significant differences in back fat depth among three sire lines at 80 kilograms and 100 kilograms liveweight but not at 120 kilograms liveweight.

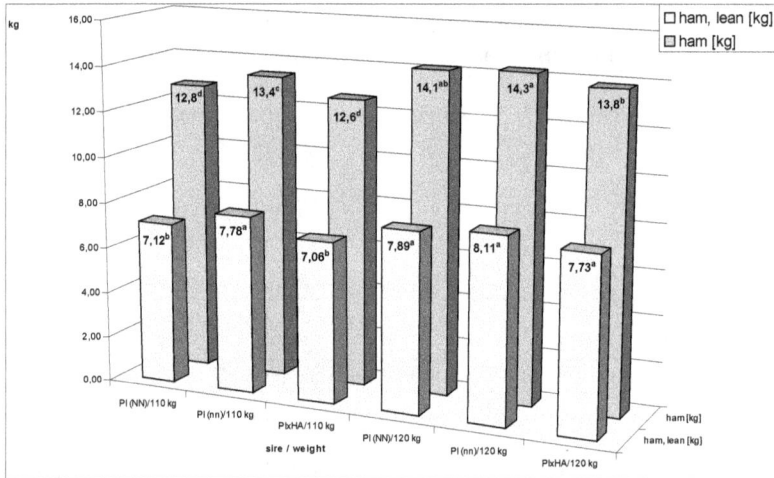

a, b, c = means with different index letter per row are significantly different (p ≤ 0.05)

Figure 2. Weights of ham and ham-lean from SIRE at 110 kg and 120 kg slaughter weight [n = 16].

References

ALZ, 1997. Richtlinie f. d. Stationsprüfung auf Mastleistung, Schlachtkörperwert und Fleischbeschaffenheit beim Schwein. Ausschuß für Leistungsprüfungen und Zuchtwertfeststellung beim Schwein; Zentralverband der Deutschen Schweine-produktion (ZDS)

GfE, 1987. Empfehlungen zur Energie- und Nährstoffversorgung der Schweine. Ausschuß der Bedarfsnormen der Gesellschaft für Ernährungsphysiologie; Frankfurt (Main); DLG-Verlag; ISBN 3-7690-0446-9

Ellis, M. & P. J. Avery, 1990. The influence of heavy slaughter weights on growth and car-cass characteristics of pigs. Anim. Prod. 50: 569

Ellis, M., A. J. Webb, P. J. Avery & I. Brown, 1996. The influence of terminal sire genotype, sex, slaughter weight, feeding regime and slaughter-house on growth performance and carcass and meat quality in pigs and on the organoleptic properties of fresh pork. Anim. Sci. 62: 521-530

Kuhn, M., L. Beesten & C. Jatsch, 1997. Zum Einfluß der Fütterungsintensität und des Ma-stendgewichts auf die Mast- und Schlachtleistung von Schweinen sowie das Fettsäuremuster der Gesamt- und Phospholipide des M. long. dorsi. 1.Mitteilung: Parameter der Mast- und Schlachtleistung, der Fleischbeschaffenheit sowie der Trockensubstanz- und Aschegehalt der Depotfette. Züchtungskde. 69: 294-306

Transfer of vitamin E supplements from feed into pig tissues

G. Flachowsky[1], H. Rosenbauer[2], A. Berk[1], H. Vemmer[1] & R. Daenicke[1]

[1] *Institute of Animal Nutrition, FAL, Bundesallee 50, D-38116 Braunschweig, Germany*
[2] *Institute of Chemistry and Physics, BAFF, E. C. Baumann Str. 20, D-95326 Kulmbach, Germany*

Summary

Five feeding experiments with growing-finishing pigs were carried out to investigate the influence of various additional vitamin E levels and application times on vitamin E content of some tissues and the vitamin E transfer into the animal body.

Vitamin E content of control diets amounted to 20 mg kg^{-1}. Additional vitamin E levels (0.5, 1.0, 1.2 g per animal and day) were given 7, 14 or 21 days before slaughtering and compared with 100 or 200 mg vitamin E per kilogram of feed during the total growing-finishing period. Additional vitamin E intake of various groups amounted to 7.0 up to 41.6 g per animal. Vitamin E content of tissues increased with higher dosage and longer application time. The vitamin E concentration in liver, muscle and backfat of the control group amounted to 4.5, 2.5 and 9.0 mg kg^{-1}. Similar vitamin E intake ($\approx$ 25 g) was achieved in pigs consuming 100 mg vitamin E per kilogram of feed or 1.2 g vitamin E per day during the last 21 days.

Finally additional vitamin E levels increase vitamin E content in foods of pig origin, but its contribution in improving the vitamin E transfer into animal body (about 1% of added vitamin E) and the vitamin E supply of humans is relatively low.

Keywords: Pigs, Vitamin E, Supplement, Pig tissues

Introduction

Supplementation of high vitamin E levels are very common in animal production in many countries. The reasons to give higher vitamin E levels ($\approx$ 100 mg vitamin E and more per kg feed) to domestic animals are the following:

- to improve animal performance
- to increase animal health
- to improve the antioxidative state of the animal
- to improve the quality of food of animal origin
- to increase the vitamin E content of food of animal origin in order to improve the vitamin E intake of man

The objective of the present report is to analyse the transfer of additional vitamin E levels into foods from the pigs.

Materials and methods

Five feeding experiments with growing-finishing pigs were carried out to investigate the influence of various vitamin E-levels and application times on vitamin E-content of some tissues and the vitamin E-transfer in animal body.

233

The feeding period occurred from 25 to 110 kilograms of body weight. Vitamin E-content of the control diet amounted to about 20 mg per kilogram of feed. Duration of vitamin E-application and vitamin E-amounts supplemented to the animals are shown in Table 1.

Further experimental details are given by Berk et al. (1998), Flachowsky et al. (1993, 1997), Gottschalk et al. (1994) and Rosenbauer et al. (1998).

Results

Additional vitamin E intake of various groups amounted to 7.0 until 41.6 g per animal (Table 1).Vitamin E content of tissues increased with higher dosage and longer application time. The vitamin E concentration in liver, muscle and backfat of control group amounted to 4.5, 2.5 and 9.0 mg kg^{-1} (average of 5 trials).

Similar vitamin E intake ($\approx$ 25 g per growing-fattening period) was achieved in pigs consuming 100 mg vitamin E per kilogram of feed or 1.2 g vitamin E per day during the last 21 days. Vitamin E concentration in body tissues was similar after both types of application. Vitamin E-transfer from feed into food of animal origin was very low (Table 1). Its contribution to improve the vitamin supply of humans could be negligent.

Table 1. Influence of vitamin E-intake on vitamin E-concentration in selected food from pigs and vitamin E-transfer from supplementation into food (average from 5 experiments).

Vitamin E-supplementation	Additional vitamin E-intake (g per animal)	Vitamin E-concentration of various tissues (Control: mg kg^{-1}, other groups: percentage of control values)			Vitamin E-transfer (% of supplement in edible tissue)	
		Liver	Muscle	Backfat	Only pork	incl. fat et al.
Control (without additional vit. E)	—	4.5	2.5	9.0	—	—
1 g (7 days)[1]	7	170	100	100	0	0.2
0.5 g (21 days)[1]	10.5	145	155	140	0.7	1.0
1 g (14 days)[1]	14	180	180	150	0.7	1.1
1 g (21 days)[1]	21	280	180	210	0.5	1.1
1.2 g (21 days)[1]	25.2	n d.	190	210	0.5	1.0
100 mg/kg[2]	21.7	n d.	205	210	0.6	1.2
200 mg/kg[2]	41.6	n d.	230	265	0.4	0.8

[1] vitamin E-amount per day given days before slaughtering,

[2] mg per kg feed mixture,

n. d. = not determined

Conclusion

Vitamin E supplementation of feed of pigs increases vitamin E-content of foods of pig origin. Similar vitamin concentrations in body tissues were measured in pigs consuming 100 mg vitamin E per kilogram of feed during the growing-finishing period or 1.2 g vitamin E per day during the last 21 fattening days.

The vitamin E-transfer from feed into food of pigs amounted to about 1 %.

Lower transfers were measured in beef ($\approx$ 0.2 %), similar values in milk and higher transfers were registered in poultry meat ($\approx$ 2 %) and eggs ($\approx$ 25 %, see Flachowsky, 1998; Flachowsky & Berk, 1998).

References

Berk, A., H. Rosenbauer, V. Mancini, H. Vemmer, G. Schaarmann & G. Flachowsky, 1998. Einfluß unterschiedlich hoher Vitamin E-Gaben an Mastschweine auf die Fleisch- und Speckqualität in Abhängigkeit von der Lagerung. Z. Ernährungswiss. 37: 171-177

Flachowsky, G., 1998. Transfer von Vitamin E-Zulagen in der Tierernährung in Lebensmittel tierischer Herkunft. Vitaminspur 13: 40-44

Flachowsky, G. & A. Berk, 1998. Carry over of additional vitamin E amounts from feeds into food of animal origin. Proc. 3rd Nutr. Symp. 15.-20. Okt. Karlsruhe, 135-142

Flachowsky, G., T. Langbein, H. Böhme, A. Schneider & K. Aulrich, 1997. Effect of false flax expeller combined with short-term vitamin E supplementation in pig feeding on the fatty acid pattern, vitamin E concentration and oxidative stability of various tissues. J. Nutr. Phys. a. Anim. Nutr.187-195

Flachowsky, G., F. Schöne, H. Graf, G. Schaarmann, C. Kinast & F. Lübbe, 1993. Einfluß zusätzlicher Vitamin E-Gaben an unterschiedlich gefütterte Mastschweine auf den Vitamin E-Gehalt in ausgewählten Körperproben und die oxidative Stabilität des Fettes. Proc. 4. Symp. "Vitamine und weitere Zusatzstoffe bei Mensch und Tier", Jena, 30.09./01.10.1993, 112-117

Gottschalk, K., C. Kinast, H. Graf, G. Schaarmann, F. Schöne, G. Flachowsky & G. Möckel, 1994. Einfluß unterschiedlich langer, zusätzlicher Vitamin E-Gaben auf den Vitamin E-Gehalt in ausgewählten Körperproben und die oxidative Stabilität des Fettes bei Mastschweinen. Proc. 3. Tagung Schweine- und Geflügelernährung, Halle, 29.11./01.12.94, 76-79

Rosenbauer, H., H. Vemmer, K. O. Honikel & G. Flachowsky, 1998. Einfluß von Dauer und Höhe der Vitamin E-Supplementierung in der Schweinemast auf den Vitamin E-Gehalt in verschiedenen Geweben und daraus hergestellten Produkten. Proc. Soc. Nutr. Physiol., 7: 134

Effect of Comfrey (symhytum peregrinum) fed to pigs on meat quality traits

G. Bee, G. J. Seewer Lötscher & P.-A. Dufey

Swiss Federal Research Station for Animal Production, 1725 Posieux, Switzerland

Summary

Comfrey is known for the high crude protein content and substances which positively affect growth performance and health status of the pig. However little is known about the influence on determinant parameters of meat quality. In the present study 22 Large White pigs were fed based on liveweight, either a common growing-finishing diet (A) or the same diet supplemented with 10% leaves from comfrey (B). The animals were slaughtered at an average liveweight of 105 kilograms. Tissue samples of backfat and m. *longissimus dorsi* were collected 24 hours after slaughter. Animals of treatment B had lower stearic and palmitic ($P<.05$), but higher oleic and linolenic acid ($P<.05$) concentration in the adipose tissue than those of treatment A. The differences between treatments were not evident in the muscle lipids. Drip and cooking losses as well as both average pH values measured 45 minutes and 24 hours *post mortem* were not affected by the comfrey supplementation. With respect to colour measurements, the a* component (redness) and chroma tended to be higher ($P<.06$) in treatment B compared to A. Furthermore, the taste panel evaluation did not reveal any treatment differences. The data of the present study suggest, that a moderate comfrey supplementation has an impact on the composition of backfat lipids without affecting other meat quality traits.

Keywords: Comfrey, Symhytum peregrinum, Meat quality, Pigs

Introduction

Comfrey could be a cheap feedstuff for pigs, producing a high yield of forage with little labour especially as Keindorf & Keindorf (1978) reported that pigs seems to like it. As reported in earlier studies, the high protein content as well as other ingredients seem to favourably affect health and growth performance of growing finishing pigs. Nakanishi *et al.* (1978) reported no negative effect of dietary comfrey supplementation up to 25% of a growing-finishing diet on growth and slaughter performance compared to a control diet. However, the roots and leaves contain substantial quantities of Pyrrolizidin alkaloids (.01-.15% DM), which are known to be hepatotoxic (Abbott, 1988). Furthermore Keindorf & Keindorf (1978) reported that comfrey harvested from May to September contains considerably high amounts of nitrate (5.4 to 32 g/kg DM), which fed in higher dosages can unfavourably affect animal health.

To date little is known on the influence of dietary comfrey supplementation on meat quality traits. Therefore the present study was carried out to evaluate the impact of a 10% comfrey supplementation on the backfat fat composition as well as on meat quality parameters.

Material and methods

Treatments

The experiment was conducted on Swiss Large White pigs (14 female and 9 castrated male pigs). At 25 kilograms liveweight, animals were randomly assigned to either a control base diet (A) or the base diet supplemented with 10% of comfrey meal (B). The feed was supplied according to a liveweight (LW) based regimen.

Sampling of meat and fat tissue

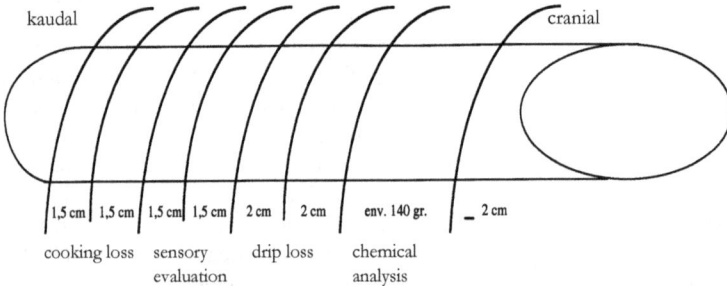

Figure 1. Sampling scheme for meat quality measurements in the longissimus dorsi muscle

The animals were slaughtered at an average LW of 105 kilograms in the slaughterhouse of the RAP. Following 24 hours of chilling the carcasses were dissected according to the guidelines of the MLP Sempach (Rebsamen *et al.*, 1995). At the height of the 9th rib pH measurements (WTW, Model pH196-S, Wintion Elektrode) in the *longissimus dorsi* muscle (LD) were assessed within 45 minutes (pH$_i$) and 24 hours (pH$_u$) *post mortem*. At the same position backfat tissue was collected for fatty acid profile analysis. LD muscle was taken from the position of the 9th rib and 20 cm backward for chemical analysis, drip loss, cooking loss and shear force measurements as well as sensory evaluation according to the scheme shown in Figure 1.

Meat quality measurements

Drip loss was measured according to standardised methods (Honikel, 1998) in 1.5 cm thick slices during 24 and 96 hours at 4°C. Samples to evaluate cooking losses were conditioned for four days at 4°C, subsequently vacuum-packed and stored at –20°C. 24 hour prior cooking samples were thawed at 4°C and subsequently cooked on a grill plate (190-195°C; Beer Grill AG, Zürich, Schweiz). Samples were weighed after each step and losses during conditioning, freezing and cooking were estimated. After cooking, samples were cooled and sheared (five individual measurements per sample) on a Warner Bratzler apparatus (Salter, Manhattan, USA). All measurements were carried out in duplicate. Colour of LD was measured using a Minolta Chroma Meter CR-300 with the light source D$_{65}$ (Minolta, Dietikon, Switzerland) to obtain L*, a*, and b* and to calculate the Chroma values.

Sensory evaluation

The eating quality was assessed by a trained sensory panel (12 panelists). The samples were prepared as loin chops and cooked as stated before. The chops were evaluated for taste intensity by a pair wise comparison. It was emphasised that the comparisons were carried out with samples from animals of the same litter, sex and slaughter date. To avoid the effect of the order (samples from treatment A vs. treatment B) during the sensory evaluation the order was inverse for half of the panelists.

Chemical analysis

Chemical analysis of the muscle (dry matter, crude ash, protein, crude fat and fatty acid profile of muscle lipid) and of the backfat tissue (dry matter and fatty acid profile) was carried out according to methods of the accredited laboratory of the Swiss Federal Station for Animal Production – (RAP, Posieux, Switzerland).

Statistical analysis

The data were analysed by one-way-ANOVA with block design (NCSS 97, 1997). The pH, drip and cooking loss data were analysed as repeated measurements. In the tables results are presented as least square means and compared by least significant difference at $P < .05$. The sensory evaluation results correspond to discrete variables and follow a binomial distribution. To assess differences between treatments a double sided test was performed.

Results and discussion

Muscle and backfat tissue composition

Dry matter content of LD from animals of treatment A was lower than treatment B, but no further effects on nutrient composition could be detected (Table 1). The muscle lipid content amounted to 2.2 to 2.4 % in the wet tissue (9.0 to 9.8 % DM) and complies with the breeding strategy of the MLP Sempach for increased intramuscular fat content to achieve good meat quality in swine. In contrast to nutrient composition, the fatty acid profile from muscle (Table 1) and backfat lipids (Table 2) were affected by the dietary treatment. In the backfat and to a lesser extent in the LD, the amount of deposited saturated fatty acids (SFA) was lower due to comfrey supplementation, whereas monounsaturated fatty acid content (MUFA) was increased in the backfat but not in the LD. The differences were mainly on account of altered deposition of palmitic (16:0), stearic (18:0) and oleic acids (18:1). It is unlikely that the changes are connected with the dietary comfrey supplementation since the fat content of the plant is low and none of the cited fatty acids are present in high quantities. One possible explanation might be the lowered body fat deposition rate due to a slightly decreased growth of animals in treatment B (data not shown). These could account for the altered SFA amount due to lowered *de novo* synthesis (Chilliard, 1993) and/or higher Δ-9 desaturase activity.

Table 1. Dry matter, nutrient composition (g/kg) and fatty acid profile of LD.

| | Treatment | | |
	A	B	P
Dry matter (%)	24.7	24.0	.035
Crude ash	43.3	43.9	.589
Crude protein	850.6	864.9	.114
Crude fat	98.2	90.4	.434
Fatty acid profile (g/kg total lipid)			
16:0	26.38	25.67	.042
18:0	13.90	13.30	.108
18:1	43.53	43.69	.787
18:2	6.96	7.78	.218
18:3	.32	.43	<.001
20:4	1.67	1.81	.541
SFA	41.97	40.57	.048
MUFA	48.13	48.20	.922
PUFA	9.89	11.17	.192

Table 2. Fatty acid profile of backfat lipids.

| | Treatment | | |
	A	B	P
Fatty acid profile (g/kg total lipid)			
16:0	25.73	25.06	.011
18:0	16.72	15.22	.001
18:1	39.47	41.59	.002
18:2	10.64	10.32	.328
18:3	.84	.96	<.001
20:4	.08	.06	.043
SFA	44.35	42.19	<.001
MUFA	42.88	45.26	.002
PUFA	12.78	12.55	.535

Meat quality traits

The meat quality traits are shown in Table 3. The pH value measured in the LD muscle 45 minutes (pH_i) and 24 hours (pH_u) *post mortem* indicate a normal pH decline during the 24 hours post slaughter. At both time points no significant treatment effects could be detected. Four and three animals respectively had 45 minutes *post mortem* pH values in the range of 5.60 and 5.79, and below 5.60, respectively which indicate an incidence of PSE. However, the occurrence was not related to one of the treatments which is in agreement with previous results (Dufey, 1998) and point out that within the same breed among lines the glycolytic potential of the LD muscle varies. The L* (lightness) and b* values were not affected by the dietary treatments. However, the a* (redness, P = .05) and Chroma values (colour saturation, P = .07) tended to be higher in the comfrey treated group. Drip loss and losses after freezing and cooking as well as shear force values were not affected by the diet. Although apparently the same method was used to assess drip loss, the present values are higher compared to other studies (Karlsson *et al.*, 1993; Maribo *et al.*, 1998), but are in agreement with results of previous investigations at our laboratory (Dufey, 1998).

The aim of the sensory analysis was to detect possible effects of comfrey on taste intensity. The results revealed no differences on eight out of the eleven pair wise comparisons. Where a difference (three pair wise comparisons) was noticed, the chops from animals in treatment A had a stronger taste. The panelists could no further describe the stronger taste. We stopped doing additional sensory analysis, since the results from the screening turned out very clear.

Table 3. Meat quality traits.

	Treatment		
	A	B	*P*
pH$_i$	5.82	5.86	
pH$_u$	5.38	5.43	.122
CIE colour			
L*	51.9	51.2	.470
a*	6.0	7.0	.050
b*	2.2	2.5	.439
Chroma value	6.4	7.4	.067
Drip loss (%)			
0 – 24 hr	7.05	8.55	
24 – 96 hr	4.84	3.48	.831
0 – 96 hr	11.97	11.94	.966
Water holding capacity (%)			
after freezing	16.97	17.78	
after cooking	18.36	18.92	.195
Total loss	32.23	33.31	.205
Shear force (kg)	3.80	3.66	.363

The present results on nutrient composition and meat quality traits demonstrate, that comfrey supplied to growing-finishing pigs at a level of 10% of the diet had only minor impacts on meat quality parameters.

References

Abbott, P. J., 1988. Comfrey: assessing the low-dose health risk. Medical Journal of Australia 149: 678-682

Chilliard, Y., 1993. Dietary fat and adipose tissue metabolism in ruminants, pigs, and rodents: a review. J. Dairy Sci. 76: 3897-3931

Dufey, P.-A., 1998. Engraissement des porcs avec un apport supplémentaire de vitamine E dans la ration. II. Effets sur les propriétés sensorielles et physicho-chimiques de la viande. Revue Suisse Agricol. 30: 186-193

Honikel, K. O., 1998. Reference methods for the assessment of physical characteristics of meat. Meat Sci. 49: 447-457

Karlsson, A., A.C. Enfalt, B. Essen-Gustavsson, K. Lundstrom, L. Rydhmer & S. Stern, 1993. Muscle histochemical and biochemical properties in relation to meat quality during selection for increased lean tissue growth rate in pigs. J.Anim.Sci. 71: 930-938

Keindorf, A. & H. J. Keindorf, 1978. Nitrate and nitrite poisoning in pigs caused by eating comfrey. Monatshefte für Veterinärmedizin 33: 425-427

Maribo, H., E.V. Olsen, P. Barton-Gade, A.J. Møller & A. Karlsson, 1998. Effect of early post-mortem cooling on temperature, pH fall and meat quality in pigs. Meat Sci. 50: 115-129

Nakanishi, G., M. Akahori, T. Ohmi & Y. Niwa, 1978. Studies on the suitability of comfrey as feed for pigs. Food Science and Technology Abstracts 35: 271-281

NCSS 97, 1997. NCSS 97, Ed. 97, Number Cruncher Statistical Systems, Kaysville, Utah, USA.

Rebsamen, A., D. Schwörer & D. Lorenz, 1995. Die Schlachtkörperzerlegung beim Schwein in der MLP Sempach. Der Kleinviehzüchter 43: 223-259

Author index

Adamec, T. 193
Agárdi, G. 217
Amarger, V. 157
Amigues, Y. 139
Andersen, H. J. 15
Andersen, S. 123
Andersson, L. 157
Barba, C. 175
Barowicz, T. 225
Baulain, U. 135, 181
Bečková, R. 151
Bee, G. 237
Berk, A. 233
Bernardo, A. 85
Bidanel, J.P. 37
Billon, Y. 139
Bocian, M. 143, 147
Brade, W. 181
Busk, H. 129
Caritez, J.C. 139
Čechová, M. 221
Chardon, P. 157
Csapó, J. 185
Csapó-Kiss, Zs. 185
Cumbreras, M. 175
Czarnecki, R. 213
Daenicke, R. 233
David, P. 151
de Vries, A.G. 27
Delgado, J.V. 175
Diestre, A. 171
Dřímalová, K. 221
Dufey, P.-A. 237
Elsen, J.M. 139
Ender, K. 73
Faucitano, L. 27
Fernandez, X. 47
Ferreira-Cardoso, J. 85
Flachowsky, G. 229, 233
Gajic, Z. 115
Geysen, D. 119
Gispert, M. 171
Gläser, K. R. 203
Glodek, P. 229
Grajewska, S. 143
Grindflek, E. 43

Gundel, J. 217
Házas, Z. 185
Henning, M. 135
Hermán Ms. A. 217
Hertel, S.H. 129
Hofer, A. 69
Horn, P. 185
Húsvéth, F. 185
Isakov, V. 115
Jacyno, E. 213
Jamison, W. 101
Janssens, S. 119
Jeon, J.T. 157
Kaczorek, S. 165
Kallweit, E. 181
Kalm, E. 157
Kapelanska, J. 207
Kapelanski, W.   111, 143, 147,
                                   189, 207
Karamucki, T. 147
Karlsson, A. 129
Karlsson, A. H. 47
Kjos, N. P. 81
Klont, R. E. 47
Koćwin-Podsiadła, M. 165
Köhler, P. 181
Kohn, G. 135
Kolstad, K. 199
Kortz, J. 143, 147
Koucký, M. 193
Kratz, R. 229
Krzęcio, E. 165
Küchenmeister, U. 73
Kuhn, G. 73
Kuryl, J. 143, 147
Lagant, H. 139
Lahucky, R. 135
Laštovková, J. 193
Le Roy, P. 139
Leroy, P. L. 161
Lien, S. 43
Liu Z. 43
Looft, C. 157
Lorenz, D. 69
Matos, J. 175
Michalska, G. 111, 189

Mikule, V. 221
Milan, D. 157
Monin, G. 139
Moreno, C. 139
Naděje, B. 193
Nowachowicz, J. 111
Nowachowicz, J. 189
Nürnberg, K. 73
Øverland, M. 81
Paul, S. 157
Pedersen, B. 123
Pietruszka, A. 213
Pires da Costa, J. S. 85
Plastow, A. 27
Prokop, V. 221
Przybylski, W. 165
Puigvert, X. 171
Rak, B. 111
Rak, B. 189
Rak, B. 207
Ramos, A.M. 175
Rangel-Figueiredo, T. 175
Rebsamen, A. 69
Reinsch, N. 157
Rey, V. 157
Robic, A. 157
Rogel-Gaillard, C. 157

Rosenbauer, H. 233
Rybarczyk, A. 143
Santos e Silva, J. 85
Scheeder, M.R.L. 203
Schulz, E. 229
Schwörer, D. 69
Seewer Lötscher, G. J. 237
Sellier, P. 139
Soler, J. 171
Sosnicki, A. 27
Szelényi Ms, M. 217
Szyda, J. 43
Talmant, A. 139
Tibau, J. 171
Tornsten, A. 157
Tribout, T. 37
Tvrdoň, Z. 221
Vandepitte, W. 119
Varga-Visi, É. 185
Vemmer, H. 233
Verleyen, V. 161
Vernin, P. 139
Wenk, C. 7, 203
Yerle, M. 157
Zimmermann, M. 93
Zurawski, H. 207
Zybert, A. 165

Subject index

Adipose tissue 203
Backcross 43
Breed 85, 171, 193
Breed differences...................................199
Breeding 27
Ca salts of fatty acids 225
Cancer..93
Carcass 115
Carcass grading 207
Carcass meatness...................................213
Carcass merits 81
Carcass quality 129
Characterisation....................................175
Cholesterol 225
Cholesterol content...............................185
Colon..93
Comfrey...237
Conservation...................................85, 175
Consumers..101
Correlation.......................................37, 123
Correlation coefficient..........................213
Crossbreds 193
Crossbreeding..111
Crosses 229
DNA markers 27, 175
Dripping water loses..............................221
Duroc..129
Eating quality 15
Energy level..115
Extrinsic 101
Fat...185
Fat deposition 199
Fat distribution 199
Fat quality...73, 203
Fat score 203
Fatteners 213
Fatty acids 73, 185, 203
Feed conversion.....................................213
Feeding..225
Food waste products................................81
Genetic 115
Genetic parameters...........................37, 119
Genotypes of RYR 151
Growing-finishing pig...........................217
Growth performance 81
Growth rate.......................................147, 213
Halothane 129, 139
Halothane genotype...............................171

Halothane locus......................................161
Hampshire..189
Heritability 37, 123
Heterosis 111
Intramuscular fat............69, 151, 181, 221
Intrinsic 101
Lean content...143
Lean growth rate 119
Lean meat...151
Major gene 27, 139
Meat 47, 93, 225
Meat colour 123
Meat percentage.....................................229
Meat quality . 27, 37, 73, 81, 111, 115, 129,
   135, 139, 143, 147, 157, 161, 165, 171,
   181, 189, 193, 207, 237
Meatiness 165
Membrane 73
MHS gene 135
Mixed inheritance model 43
MR spectroscopy 135
Muscle fibre 47
Near infrared spectroscopy 181
Nitrogen deposition 221
pH..123, 151
pH_1 221
Piétrain 119, 129, 161, 229
Pig fattening...229
Pig tissues 233
Pigs . 27, 37, 47, 69, 73, 81, 111, 115, 135,
   139, 143, 147, 157, 171, 181, 185, 189,
   193, 203, 207, 225, 233, 237
Pork...47, 101
Pork quality...15
Pork quality characteristics.....................15
Primal cuts 229
Production traits.....................................123
PSS...175
PUFA 225
QTL detection 43
Quality 47, 85
Recovery 85
Reml 119
RN gene 139, 157, 165
RYR1 genotype 143, 147
Saturated fat 93
Selection progress...................................69
Sensorial evaluation 193

Sex 229
Slaughter weight 171, 229
Soybean oil 229
Soybean seed 217
Stomach 93
Stress 135
Stress negative 161
Sunflower cake 217
Sunflower seed 217
Supplement 233
Swine 119

Swine meat quality 43
Swiss Pig Performance Testing Station .. 69
Symhytum peregrinum 237
Technological quality 15
Technological yield................................ 165
Unsaturated fatty acids......................... 217
Values 101
Variability 85
Various genotypes................................. 185
Vitamin E.. 233

www.ingramcontent.com/pod-product-compliance
Lightning Source LLC
Chambersburg PA
CBHW061405210326
41598CB00035B/6102